# Troubleshooting & repairing VCRs

# Troubleshooting & repairing VCRs

### 3rd edition

# Gordon McComb

## TAB Books
### Division of McGraw-Hill

New York  San Francisco  Washington, D.C.  Auckland  Bogotá  Caracas  Lisbon  London  Madrid
Mexico City  Milan  Montreal  New Delhi  San Juan  Singapore  Sydney  Tokyo  Toronto

Published by TAB Books, a division of McGraw-Hill, Inc.

pbk    2 3 4 5 6 7 8  9 FGR/FGR 9 9 8 7
hc     3 4 5 6 7 8 9 10 FGR/FGR 9 9 8 7 6

**Library of Congress Cataloging-in-Publication Data**
McComb, Gordon.
     Troubleshooting and repairing VCRs / Gordon McComb.—3rd ed.
        p.    cm.
     Includes bibliographical references and index.
     ISBN 0-07-155016-X     ISBN 0-07-155017-8 (pbk.)
     1. Videocassette recorders—Maintenance and repair—Handbooks, manuals, etc.  I. Title.
     TK6655.V5M29  1995
     621.388'337—dc20                                    95-3221
                                                         CIP

Acquisitions editor: Roland S. Phelps
Editorial team: Joanne Slike, Executive Editor
               Lori Flaherty, Managing Editor
               Bert J. Petersen, Book Editor
Production team: Katherine G. Brown, Director
               Lisa M. Mellott, Coding
               Ollie Harmon, Coding
               Rose McFarland, Desktop Operator
               Nancy K. Mickley, Proofreading
               Lorie L. White, Proofreading
               Jodi Tyler, Indexer
Design team: Jaclyn J. Boone, Designer                    EL3
             Katherine Stefanski, Associate Designer      1550178

To my father, E.W. McComb,
who always put back together
what I took apart.

# Contents

Acknowledgments   xiii

Introduction   xv

## 1 Introduction to videocassette recorders   1

VCR formats   2
VCR types   5
Video and audio heads   6
Tuning   11
Operating controls   13
Basic operation   14
The enhanced VCR   15

## 2 How VCRs work   23

Television history 101   23
What is television?   25
Baseband and RF signals   29
Basic VCR block diagram   30
Cassette loading   31
Tape threading   32
Tape transport   37
Recording and playback operations   41
Video head geometries   48
Inside hi-fi audio   55
Super Beta   59
Super VHS   59
Digital effects   62
Time-base corrector   64
Elapsed-index or real-time counter   67

Auto tracking  68
Hi-band 8mm  68
Fuzzy logic  68
PCM audio  69
Cable box control  70
Controls  70
Connectors  71

## 3 The VCR environment  75

Unpacking  75
Installation  76
Hookup  77
Checkout  86
If something goes wrong  87
Achieving optimum picture and sound  87
Proper tape handling  90
Video accessories  93
Turbocharging your VCR  99
Setting up a home theatre  107

## 4 Tools and supplies for VCR maintenance  109

Workspace area  109
Basic tools  110
Volt-ohmmeter  110
Logic probe  114
Logic pulser  117
Oscilloscope  118
Frequency meter  119
Infrared detector  120
TV set or monitor  125
Assorted supplies  125
Capstan roller rejuvenator lathe  129
Using test tapes  130

## 5 Cleaning and preventive maintenance  133

Frequency of checkup  133
A word of caution  135
Personal safety  135
General maintenance  136

Internal preventive maintenance   139
Video heads   152
Oiling and lubricating   159
Tuner cleaning   161
Demagnetizing   162
End-of-tape sensor cleaning   162
Reel spindles   163
Front-panel controls   165
Final inspection and reassembly   166
Checkout   167
Maintenance log   167
Remote cleaning   167
Tape troubles   171

6 VCR first aid   183

Tools and safety   183
Dropped deck   183
Fire damage   187
Water damage   189
Sand, dirt, and dust   191
Foreign objects   192
Leaked batteries   193
Broken remote   194
Damaged tapes   195

7 Troubleshooting and repairing non-VCR problems   197

Start at the TV   197
Program reception   200
Interference   201
VCR cables and controls   211
Tracking control   211
TV/Video switch   212
Remote interference   213
Using a known, good tape   214
Bad tape rewinder   216
Macrovision mayhem   217
Troubleshooting checklist   225

# 8 Troubleshooting techniques and procedures 227

The essence of troubleshooting 227
Troubleshooting flowcharts 229
Using the troubleshooting charts in chapter 9 229
What you can and can't repair 231
What to do if you can't repair it 232
Finding a trustworthy repair technician 233
Troubleshooting techniques 234
Going beyond the flowcharts 237

# 9 Troubleshooting VCR malfunctions 239

Repair it yourself 239
Service politics 240
Maintenance procedures and special adjustments 241
Using the troubleshooting flowcharts 247
Flowchart 1. VCR does not turn on 249
Flowchart 2. VCR turns on but nothing else 255
Flowchart 3. Cassette will not load 260
Flowchart 4. Cassette will not eject 265
Flowchart 5. VCR will not thread tape 268
Flowchart 6. VCR will not play tape 270
Flowchart 7. VCR eats tape 273
Flowchart 8. Fast forward or rewind won't operate 279
Flowchart 9. Search does not work correctly 281
Flowchart 10. Tracking control has no effect 283
Flowchart 11. VCR does not respond
  to some or all front-panel controls 285
Flowchart 12. Front-panel indicators not functioning 288
Flowchart 13. Timer does not operate properly 290
Flowchart 14. Tuner channels don't change on TV 292
Flowchart 15. Sound okay; video not okay 294
Flowchart 16. Video okay; sound not okay 297
Flowchart 17. Snowy video; poor audio 299
Flowchart 18. Remote control does not
  operate or function properly 302
Flowchart 19. Unusual mechanical noises
  during loading and playback 305
Flowchart 20. You receive an
  electrical shock when you touch the VCR 307
Flowchart 21. VCR overheats 310
Miscellaneous VCR difficulties 312

## 10 Repairing camcorders 321

Disassembly 321
Head cleaning 322
Dusting and general cleaning 323
Parts cleaning and replacement 323
Camera care 324
How to help your camcorder live longer 325
Camcorder care and feeding 325
Don't forget the supplies and accessories 326
Proper care and use of camcorder batteries 326
Special tips when traveling with a camcorder 327

## Appendices

A Sources 329
B Further reading 337
C Soldering tips and techniques 341
D Attaching an F connector 345
E Television frequency spectrum 349
F Who makes what 355

Glossary 361
Index 373

# Acknowledgments

My father, Wally McComb, helped immensely with the research of the first edition of this book; his efforts and insight are much appreciated. I also want to thank Southwest General Industries for providing me with their service records and expertise and Roderick Woodcock and Marc Wielage for loaning me a portion of their encyclopedic knowledge of video. Gratitude is also due to the folks on the CEVIDEO forum of CompuServe who, though they might not have known it, contributed to the technology of this book: Bill Rood, Kaon Koo, Todd Handley, Ed Ellers, and many others. Thanks also go to Stan Pinkwas of *Video Magazine* for helping me promote the previous editions of this book and for allowing me to contribute articles to the magazine.

My wife, Jennifer, helped me through the rough spots and saw to it that I didn't miss my deadlines . . . much. Well, at least she tried. Finally, a warm thank you for the patience of my agent, Bill Gladstone, and to Roland Phelps and the rest of the gang at TAB/McGraw-Hill.

# Introduction

Have you ever wished you could freeze time or turn back the clock to a happier moment? What if you could edit out the boring parts of your life? You can do all this and more without strange science-fiction gadgetry, haunting sorcery, or medieval magic. All you need is a VCR. VCRs allow you to become a programming executive. You can record programs off the air for later viewing, tape important moments of time with the help of a video camera, and watch full-length movies in the comfort of your own living room.

Home VCRs were introduced in 1976 and consumers have bought more than 100 million of them since. More than 12 million VCRs are sold each year. More than 80 percent of all households in the United States have at least one VCR, and many have two or more. VCRs have evolved from being considered toys for yuppies to become essential appliances, as necessary for existence in this technological age as a refrigerator or oven.

But VCRs are machines after all, and therein lies a problem. Machines need care, and they break down now and then. The cost of sending a VCR to a video doctor can be astronomical—as high as $100 for a simple belt change or adjustment. Cleaning and lubricating a VCR, a job that should be done every 12 to 18 months, often costs more than $100. Consequently, owning a VCR can be an expensive proposition.

Though a VCR might be a bundle of high-tech circuitry, servo motors, and precision parts, keeping one in good health needn't be a complicated matter. In fact, taking care of a VCR isn't any more difficult than taking care of your car. Maintaining and repairing a VCR takes but a small assortment of inexpensive tools and the right instructions. The tools can be purchased at almost any hardware or electronics store, and the "right instructions" are contained in this book, which is designed expressly for you—the consumer.

This book offers step-by-step details on the care and feeding of home VCRs, from simple parts cleaning and lubrication to troubleshooting power supplies and logic circuitry problems. It is the first book to cover in an in-depth but nontechnical manner the care and repair of all types of home videocassette recorders, including Beta, VHS, and 8mm (millimeter), as well as the ever-popular *camcorders* (combination cameras and portable videocassette recorders).

The instructions in this book are simple and straightforward. Complex theory is kept to a minimum. Instead, there's a liberal blend of actual hands-on help, tips, and techniques—the type of information you most need. This book is designed and written for you as a video hobbyist or electronics enthusiast. It is technical in nature but does not require an intimate knowledge of video, audio, computer technology, electronics, or maintenance and repair techniques. You'll find plenty of introductory information plus tips to help you enjoy your VCR investment and save time and money when (and if) repairs become necessary.

About 90 percent of all VCR malfunctions are mechanical, so much of this book discusses mechanical breakdown—how to avoid it and how to repair it. That doesn't mean the book lacks information on troubleshooting (and in some cases repairing) VCR circuitry. On the contrary, every major electronic subsystem of VCRs is fully covered, including power supplies, solenoid controls, motors, and more.

All home VCRs work on the same basic principles, but specific maintenance and repair techniques vary widely among models. Therefore, this book contains specifications and repair notes for more than four dozen different video recorders and includes a handy cross-reference chart on models that are sold by many different firms but are made by only a handful of companies.

Videocassette recorders are loaded with specialty parts like polished video heads, complex threading mechanisms, and proprietary surface-mount integrated circuits. You can't get these things at your local Radio Shack and most manufacturers don't sell replacement parts to consumers. Even if you could get the components, they require special alignment tools and test jigs to properly test and install. All this means that if your VCR has a serious problem, you, as a home-based technician, can do little to make the repairs yourself. Unless you have the specific knowledge of servicing your particular brand of player, an oscilloscope to diagnose waveforms, all the specialty tools on hand, and a service manual, it is better to have serious ailments serviced at a repair center. You are free to attempt larger scale repairs, of course, but they are beyond the scope of this book.

Fortunately, malfunctions in crucial components of VCRs are rare, even with the inexpensive models. Most problems are caused by dirty switch contacts, broken wires, dirty video heads, old and worn rubber belts and rollers, and damaged tapes. In fact, these problems represent the greatest percentage of service calls to repair centers. Fortunately, you can correct these faults yourself with a minimum of tools and time. This book shows you how.

You can minimize repairs—whether done by you or someone else—by keeping your VCR in tip-top shape. This book presents an easy-to-follow preventive maintenance schedule that you can use to keep your deck working at its highest capacity. Even if you can't repair the VCR yourself, this book serves another important purpose: it helps you to be well informed about the possible causes of videocassette recorder problems. You will be better able to describe your deck's illness to the repair technician, and by specifically stating what is wrong, you have a greater chance that the problem will be fixed correctly the first time, and at a lower cost.

You also will be in a better position to spot unscrupulous repair tactics, like charging for parts that were never replaced or labor above and beyond what was needed to service a component. Most service centers and repair technicians are honest and fair, but there are exceptions. Be on the lookout for them!

Here is a brief rundown of what you'll find in each chapter.

☐ Chapter 1, "Introduction to videocassette recorders," covers the basics of video and videocassette operation, an introduction to tape formats (VHS, Beta, 8mm), and the enhanced (fully loaded) VCR.

☐ Chapter 2, "How VCRs work," includes an in-depth look at how VCRs record and play back tapes; differences between VHS, Beta, and 8mm formats; all about tape transports; inside hi-fi audio 8mm sound; and operating controls.

☐ Chapter 3, "The VCR environment," covers setting up a VCR in your home; avoiding problems in installation; how to hook up a VCR to a TV, hi-fi, antenna, and other components to enhance the picture and sound; proper tape handling; and video accessories.

☐ Chapter 4, "Tools and supplies for VCR maintenance," tells you about workspace area, basic tools for disassembly, volt-ohmmeters and their use, logic problems, scopes and frequency counters, assorted cleaning supplies, lubricating oils and greases, building a sensor for infrared light, making a capstan roller rejuvenator lathe, and using test tapes.

☐ Chapter 5, "General cleaning and preventive maintenance," covers preventive maintenance schedules and procedures for cleaning the video heads, interior cleaning, lubricating, checkout, tape care and repair, and cleaning the remote-control unit.

☐ In chapter 6, "VCR first aid," you'll find emergency procedures for minimizing or eliminating damage caused by fire, water, dirt, sand, foreign objects, and other contaminants.

☐ Chapter 7, "Troubleshooting and repairing non-VCR problems," shows you how to identify and correct video and audio problems that are not caused by the VCR, including bad tapes, improper hookup, Macrovision anticopy signals, and a maladjusted or malfunctioning TV.

☐ In chapter 8, "Troubleshooting techniques and procedures," you'll learn a logical approach to VCR repair, how to analyze problems, how to use the flowcharts in chapter 9, proper troubleshooting etiquette, how to determine what you can and can't fix, and what to do if you can't fix a problem.

☐ Chapter 9, "Troubleshooting charts for VCR malfunctions," includes flowcharts for a variety of common ailments, including decks that don't turn on, won't play tapes, only work for a few moments then stop, ignore front-panel controls, and more.

☐ Chapter 10, "Repairing camcorders," covers routine camcorder maintenance and cleaning, how to safely clean camcorder heads, dusting and general cleaning, parts cleaning and replacements, and camera care.

☐ Appendix A contains names and addresses of VCR manufacturers and makers of video-repair products.

☐ Appendix B suggests further reading for those of you who would like to broaden their knowledge of video and VCRs.

☐ Appendix C contains soldering tips and techniques. If you have not done much soldering, refer to appendix C for helpful instructions on how to solder and desolder components and wires.

☐ Appendix D explains how to attach an F connector to a 75-ohm coaxial cable.

☐ Appendix E contains the television frequency spectrum. Refer to this appendix for a chart of the TV frequency spectrum and how the frequency is broken down into channels.

☐ Appendix F will help you find out who makes what. It includes a reference list of popular VCR brands and who makes them.

# Introduction to videocassette recorders

*1*

TO APPRECIATE THE ADVANCES THAT HAVE BEEN MADE IN video, suppose you could turn the clock back in time. Imagine that you have found a time machine in your basement. You step inside the machine and go back to May 20, 1966. You arrive at 7:30 in the evening, in time to watch the evening's fare of television.

One of the first things you might notice is that although the channels are full of programming, you're limited to what you can view and when. Tune to NBC and you get the "Green Hornet." You're not a Bruce Lee fan, so you spin the dial and find "The Wild, Wild West" on CBS and "Tarzan" on NBC. Still not right. Your only other choice is to watch one of the 1940s war movies, which seem to be in steady supply on many of the local TV stations. After the evening is over, you realize that you enjoyed living in the present much better because you have control over your television viewing schedule.

You zap yourself back to the present where you are not limited to the programming tastes of television network executives. Thanks to videocassette recorders (VCRs), you can be your own programming executive—you set your own prime-time schedule to match your needs and viewing tastes.

VCRs come in many different shapes and sizes. Take a close look at VCRs and what the various features and functions do. This book investigates the three formats of VCRs and the types of features found in all models. This chapter gives an overview of videocassette recorders. Chapter 2 details how video recorders work and provides a broad overview of the science and technology found in the typical VCR.

# VCR formats

All VCRs work about the same. They record sound and picture information on magnetic tape, and then they play that information back on your TV. VCRs might be based on the same technology, but inside they are worlds apart. The first major difference between models is that there are three popular formats: VHS, 8mm, and Beta. An example of a VHS deck is shown in Fig. 1-1.

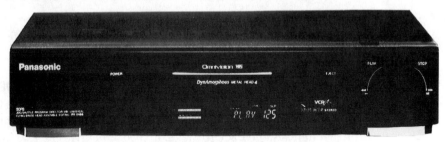

■ **1-1** *One of several hundred models of VCRs to come out since 1976.*

## Inside the three formats

Beta is the brainchild of the electronics giant Sony. Though it was not the first consumer videotape format to come out, it was the first to catch on. Sony's first Beta deck was introduced in 1975 and became widely available in 1976. Since then, Sony and a handful of other manufacturers introduced several dozen models of Beta VCRs. Beta tapes are ½ inch wide and are enclosed in a plastic cassette. The cassette protects the delicate tape from dust, dirt, and other contaminants.

A few years later, another Japanese electronics firm, JVC, introduced a rival VCR tape format. They dubbed the newcomer *VHS*; the acronym stands for *video home system*. Like Beta, VHS tapes are ½ inch wide and are enclosed in a plastic cassette. However, the cassette is physically larger, as shown in Fig. 1-2, and the signals are recorded on the tape in a different fashion than Beta. As a result, VHS tapes won't play on Beta decks and vice versa.

Through the years, the VHS format has achieved the most success, and Sony is the only remaining maker of Beta decks. Even Sony has curtailed manufacture and marketing of Beta products. Only a small handful of Beta decks are available, and most of these are high-end, semiprofessional models.

In 1982, Sony and 127 other manufacturers around the world joined forces to create a new VCR format using tape that is 8 mil-

■ **1-2** *Beta (foreground) and VHS cassettes.*

limeters wide. The new format is called *8mm* and is the most popular with *camcorders* (combination cameras and recorders). Tape cassettes for 8mm decks are only slightly larger than audio cassettes (see Fig. 1-3 on the next page), so 8mm VCRs can be made very small.

## Playback speeds

The first VCRs in all three formats recorded and played back at one speed only. This speed allowed a maximum of one or two hours of recording time on a single cassette. Rather than try to pack four or six hours worth of tape in a cassette—which is difficult at best—VCR manufacturers decided to enable alteration of the speed of the tape traveling through the deck.

Most VHS decks have three record/playback speeds: *SP* (short play or standard play), *LP* (long play), and *EP* (extended play). EP also is called *SLP* or super long play on some models. Home Beta decks can record and play back in two speeds, BII and BIII (the fast BI record speed is reserved for industrial and special semiprofessional decks only, but some Beta decks can play back BI

■ **1-3** *Audio cassette tapes are roughly the same size as 8mm cassettes.*

tapes). The 8mm decks have two speeds, SP and EP. Table 1-1 shows how the recording times measure up at the various speeds, based on commonly available tapes.

■ **Table 1-1 Recording speeds.**

| VHS | SP | LP | EP (SLP) |
|---|---|---|---|
| T-60 | 1 hr | 2 hrs | 3 hrs |
| T-120 | 2 hrs | 4 hrs | 6 hrs |
| T-160 | 2 hrs 40 min | 5 hrs 20 min | 8 hrs |

| Beta | BI | BII | BIII |
|---|---|---|---|
| L-125 | 15 min | 30 min | 45 min |
| L-250 | 30 min | 1 hr | 1 hr 30 min |
| L-500 | 1 hr | 2 hrs | 3 hrs |
| L-750 | 1 hr 30 min | 3 hrs | 4 hrs 30 min |
| L-830 | N/A | 3 hrs 30 min | 5 hrs |

| 8mm | SP | LP |
|---|---|---|
| MP-30 | 30 min | 1 hr |
| MP-60 | 1 hr | 2 hrs |

Slowing the tape decreases the quality of the picture and sound. The best quality is obtained by recording at the fast speed, as explained in chapter 2, "How VCRs work."

## VCR types

Almost all VCRs are the tabletop variety; they are designed for in-home use and plug into an electrical power outlet. There are several other subtypes available.

### Camcorders

Portable VCRs were popular until about 1985. These devices operated from battery power and, when connected to a video camera, allowed you to shoot your own tapes anywhere. Portables are no longer manufactured; they have been replaced by a combination VCR and camera called a *camcorder* (see Fig. 1-4). Camcorders are especially useful if you do a lot of away-from-home shooting because you're not encumbered by a bulky VCR, camera, and cables. Everything is self-contained. Some cam-

■ **1-4** *A camcorder—combination video camera and recorder in one unit.*

corders are record-only (you must use a tabletop VCR to watch the tape), but the vast majority have both record and playback capability.

If you have a video camera, you can use it with either a portable or tabletop model. The majority of cameras are made for portable VCRs and attach to the deck with a specialized 10- or 14-pin connector (10-pin is standard; the 14-pin type is found on some older Beta decks). Adapters are available to match the camera to the type and model of VCR. More information on camcorders is contained in chapter 10.

## Videocassette players

*VCPs* (videocassette players) lack a recording capability—they only play back tapes and lack the means to record. This means VCPs dispense with a lot of extraneous electronics, as well as the programmable record timer. As a result, VCPs are a little less expensive than full-fledged recording VCRs. However, they require the same level of mechanical maintenance as full-fledged recording VCRs.

## Video and audio heads

VCRs record and play back audio and video signals using magnetic heads. At a minimum, all VCRs use the following magnetic heads:

☐ A stationary full-erase head
☐ A stationary audio head
☐ Rotating video heads
☐ A stationary control-track head

The way these heads record information on the tape is similar for the three formats, though the exact track widths and specifications vary. The recording formats for VHS, Beta, and 8mm are shown in Fig. 1-5. The relative location of the heads in the types of VCRs is shown in Fig. 1-6.

### Full-erase head

The full-erase head erases the entire tape and is on only during recording. The erase head blanks out any previous recordings that might be on the tape. In some VCRs, a secondary audio erase head can be used. This head erases just the audio track. For more details, see the section titled "Additional heads."

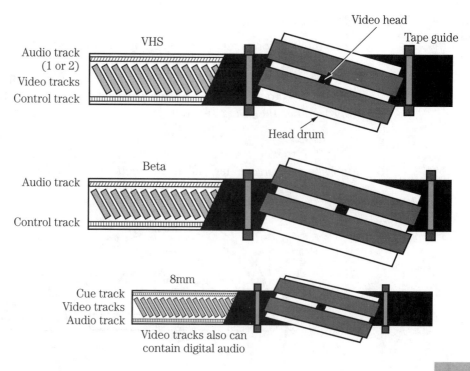

**■ 1-5** *The three home video formats and how the information tracks are recorded on each one.*

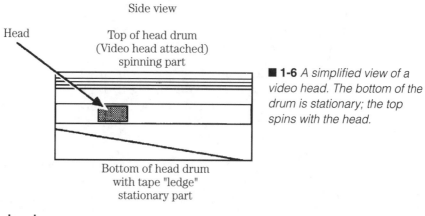

Side view

**■ 1-6** *A simplified view of a video head. The bottom of the drum is stationary; the top spins with the head.*

## Audio head

In the record mode, the tape moves past the audio head, where the head impresses the sound signal on a thin track on the edge of the tape. In the playback mode, the head picks up the signal recorded on the tape.

## Video heads

Video information is much more complex than audio information. More tape area is needed to record a picture. To pack a video picture on a ½-inch or 8mm tape using conventional recording techniques, the tape must be shuttled through the deck at speeds approaching 20 feet per second!

Obviously this is impractical, so VCRs use rotating heads, as shown in Fig. 1-7, to record and play back video. The heads are mounted in a polished metal cylinder, called the *head drum* (other names are used as well, including *scanner*, *head cylinder*, and *head wheel*).

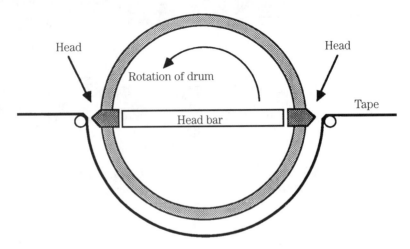

■ **1-7** *Top of a video head drum, showing both heads and the 180 degree path of the tape.*

The heads are at an angle, so they record a series of long, diagonal tracks on the tape, as shown in Fig. 1-8. The tape might creep slowly through the VCR, but the rotating video heads spin at 1800 revolutions per minute, effectively covering about 250 inches of tape each second.

Table 1-2 shows the relative speeds of the tape traveling through a VHS VCR versus *writing speed*, which is the amount of tape covered by the video heads in one second.

All VCRs (except certain Beta and VHS camcorders) use a minimum of two video heads mounted 180° apart. Only one head touches the tape at any time. The more advanced VCRs use even more video heads for improved playback at all speeds plus high-quality special effects. See chapter 2, "How VCRs work," for a more detailed discussion of what each of the video heads do.

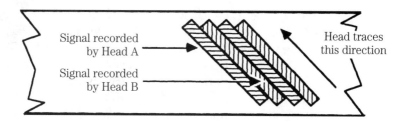

Signal recorded by Head A →

Signal recorded by Head B →

Head traces this direction ←

■ **1-8** *The heads on the head drum record adjacent tracks on the tape.*

■ **Table 1-2 Writing speeds.**

| Format and Speed | Writing Speed | Linear Tape Speed |
|---|---|---|
| **Beta** | | |
| B-I | 274.6 ips (6975mm/s) | 1.57 ips (40mm/s) |
| B-II | 275.4 ips (6995mm/s) | 0.78 ips (20mm/s) |
| B-III | 275.6 ips (7002mm/s) | 0.52 ips (13.3mm/s) |
| | | |
| **VHS** | | |
| SP | 228.5 ips (5804mm/s) | 1.31 ips (33.35mm/s) |
| LP | 229.1 ips (5820mm/s) | 0.66 ips (16.7mm/s) |
| EP | 229.3 ips (5826mm/s) | 0.44 ips (11.12mm/s) |
| | | |
| **8mm** | | |
| SP | 147.7 ips (3751mm/s) | 0.56 ips (14.3mm/s) |
| LP | 148 ips (3758mm/s) | 0.28 ips (7.2mm/s) |

Note: Writing speed increased slightly at slower linear tape traveling speeds.

## Control-track head

During recording, the control track records a series of 30 Hz (hertz) pulses. These pulses are used to synchronize the video heads during playback so they pass directly over the tracks that were previously recorded. Without the control track, the video heads might not scan directly over the video tracks, and the picture would be garbled. The control track serves the same general purpose as sprocket holes in movie film. The sprocket holes help align each frame so you see a steady picture on the screen. In almost all VCRs, the control-track head is mounted in the same unit as the audio head.

## Additional heads

Some VCR models incorporate additional heads. Their functions are discussed in the following paragraphs.

### Stereo audio heads

Monophonic VCRs use one audio head to record and play back sound. Older stereo VHS decks use a dual audio head for recording the right and left tracks separately. (Note: stationary stereo audio heads are seldom found on home VCRs these days). Audio recording with two audio heads is called *linear stereo*. The tracks are placed alongside one another near the edge of the tape (see Fig. 1-9) and occupy the same space as the standard mono audio track (monophonic decks pick up both the right and left channels together). Noise reduction circuitry (such as Dolby) is often used with linear stereo decks to improve the sound quality.

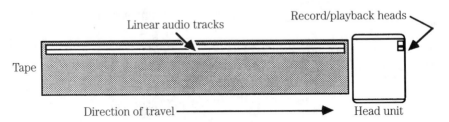

■ **1-9** *A linear stereo head records the right and left soundtracks separately at the top of the tape.*

On a number of linear stereo VCRs, one or both of the channels can be independently erased or activated for audio dubbing. The audio erase head is built into the audio head, as shown in Fig. 1-10. With audio dubbing, you record only the audio segment, but leave the video and control track intact. Most linear stereo decks let you record on either the left or right stereo channel, leaving the other channel untouched. Beta VCRs record stereo audio mixed with the picture information as explained in the following.

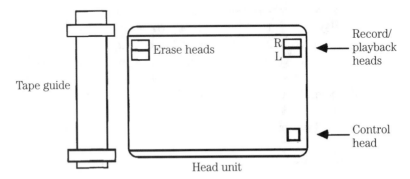

■ **1-10** *Closeup of the control/audio head unit. The unit contains both the control and audio heads and can also contain one or more audio erase heads.*

### Rotating audio heads

VHS hi-fi decks—those that record stereo sound in high fidelity—have two additional heads mounted with the rotating video heads. These heads record audio information in an *FM* (frequency modulation) carrier signal. Even though the audio and video tracks are intermixed, they do not interfere with one another because the angles (or *azimuth*) of the heads are different, as depicted in Fig. 1-11. Magnetic heads do not pick up information that is recorded on substantially different angles. The video heads themselves also record at different azimuth angles, as discussed more fully in chapter 2, "How VCRs work."

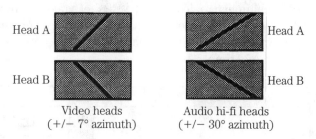

Head A

Head A

Head B

Head B

Video heads
(+/− 7° azimuth)

Audio hi-fi heads
(+/− 30° azimuth)

■ **1-11** *The video heads are slanted at an angle to prevent interference from adjacent signal tracks.*

Beta stereo hi-fi decks record picture and high-fidelity sound using the machine's standard complement of video heads. Both the sound and picture are combined in one signal and placed on the tape together.

### Flying erase head

The main full-erase head erases the contents of the entire tape prior to recording. The 8mm tabletop VCRs, top-end VHS and Beta decks, and many 8mm and VHS camcorders use one (and sometimes two) flying erase heads mounted with the video heads. The flying erase head(s) erases just the video tracks, leaving the control and audio tracks alone. A flying erase head provides better, cleaner edits.

## Tuning

VCRs are designed so that you can record programming from a TV broadcast antenna or from cable. Like your TV, VCRs incorporate tuners, so you can select the channel you want to record.

Older VCRs used mechanical-detent tuners, the same as the old-style TV sets. With mechanical-detent tuners, you cannot change the channel under program or remote control, and these tuners usually require periodic cleaning and can go out of alignment. Dirty contacts in the tuner can cause a variety of problems such as loss of signal quality and noisy reception. Other chapters show you how to clean tuner contacts.

All current VCRs come with electronic push-button tuning, as shown in Fig. 1-12. You can access *VHF* (very high frequency) channels 2 through 13 as well as *UHF* (ultra high frequency) channels 14 through 69 (a few models of modern VCRs can tune to UHF channel 83). The latest VCRs let you tune to any number of these channels. Older decks required you to preset the channels you wanted to watch. You had the ability to preset 12 to 24 channels at one time.

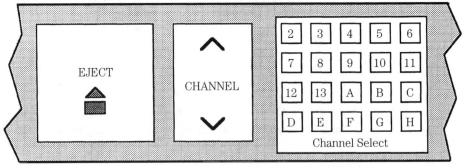

Front of VCR

■ **1-12** *A typical electronic tuner on the front of a VCR. "Dial" the channel by pressing a channel up/down button or a button on a keypad.*

The push buttons used in an electronic tuner also might need occasional topical cleaning as covered in other chapters, but they aren't as susceptible to the same trouble that plagues mechanical tuners.

## Cable ready

Cable systems that carry more than 12 channels usually broadcast the extra channels on special cable frequencies that normal TVs and VCRs can't receive. However, a cable-ready VCR (or TV for that matter) can tune to these channels. The number of accessible channels varies among VCRs but is usually from 102 to more than 140. Frequency allotments for regular and cable channels are provided in appendix E.

Having a cable-ready VCR means you don't need to use the cable company's converter box to watch the nonscrambled channels. Keep in mind that a cable-ready VCR or TV will not decode a scrambled picture. You'll still need a cable TV decoder box for this.

### Programmable VCRs

One of the benefits of owning a VCR is being able to record programs while you're away from home. Nearly all VCRs have at least a rudimentary timer built in (the timer was a separate component on the first VCRs). You select the channel you want, then set the timer to start recording at a particular hour (and sometimes day). You can program the timer to record for as little as a minute or as long as you want, up to the maximum length of the tape. This type of recording is called *time shifting*.

All modern VCRs allow you to record several different programs, at different times of the day, on different days of the week, and on other channels. You could, for instance, record the news from 8:00 to 8:30 in the morning, Perry Mason at noon, and cartoons for the kids from 2:00 to 3:30.

VCR timers are rated by the number of days and events they can be programmed for. VCRs typically have timers that allow recording of up to 7 or 14 programs (called *events*) over a one- to three-week period. The only limitation is that you can't capture more programming than the tape will hold.

**13**

## Operating controls

The earliest VCRs used clunky mechanical controls for their various operating modes such as PLAY, REWIND, and FAST FORWARD. You pushed a button that turned a lever, that rotated a crank, that flipped a ratchet, that pushed down a spring, that moved a belt, and so forth. Fully mechanical VCRs are no longer made, but if you own one, you will need to service it occasionally to keep the springs, ratchets, belts, and other components operating smoothly. Chapter 5 deals with these maintenance procedures.

Now, all VCRs use touch-sensitive electronic or electromagnetic controls. When you push a button, an electrical signal is sent to a motor or solenoid, bypassing all the intermediate mechanical linkages. Note that the machine is less complex mechanically, but a simple break in the signal lines from the switch to the actuating motor or solenoid could mean an inoperative deck.

### Remote control

Remote controls come in two forms: wired and infrared. Wired remotes are tethered to the VCR with a cable and are usually limited in function. You can stop and start the VCR but little else. Infrared remotes, which work by sending pulses of invisible light to your VCR, incorporate all or nearly all of the functions available at the deck itself.

Many of the early VCRs (before 1983 or 1984) used wired remotes, but now VCRs use infrared wireless units. Maintenance and service procedures for both types of remotes are provided in chapters 5 and 9.

## Basic operation

All VCRs have the features discussed in the following sections.

### Tape loading

Tape loading is done through manual or automatic means. You will find manual tape loading on all top-loading VCRs (no longer made except for camcorders). With manual loading, you insert a tape into the loading mechanism and push the mechanism down into play position. Automatic tape loading is found on all front-loading VCRs. With automatic loading, you insert a cassette into a loading door. The VCR grabs the tape and pulls it inside the machine. Figure 1-13 shows a top-loading VCR, and Fig. 1-14 shows a front-loading VCR.

### Function controls

The basic operator controls are PLAY, RECORD, FAST FORWARD, REWIND, and PAUSE. The PLAY button plays a previously recorded

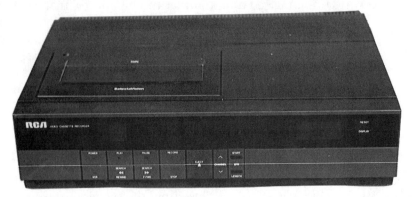

■ **1-13** *An older top-loading VCR.*

■ **1-14** *A front-loading VCR.*

tape, and the RECORD button makes a new tape. The FAST-FORWARD and REWIND buttons shuttle tape back and forth within the cassette, and the PAUSE button temporarily stops the VCR but does not fully disengage recording or playback.

### Input/output terminals

The input and output terminals serve as gateways in and out of the VCR. The input terminals provide a signal for the VCR to record. Conversely, the output terminals route the playback signals from the VCR to a television set.

Two types of terminals are used: *RF* (radio frequency) and baseband video/audio. RF signals combine both audio and video on a modulated carrier and are broadcast on a television channel, usually channel 3. *Baseband* signals consist of only raw audio and video and are used to connect to a monitor or other VCR. For more details on baseband and RF signals, see chapter 2.

## The enhanced VCR

There are plenty of special features available on VCRs. Following is a summary of the enhancements you might see on decks in almost all price ranges.

### Special video effects

If you're a TV sports nut or have ever even watched a football game on TV, you've seen special video effects at work. Many home VCRs are capable of the special effects used in sports broadcasts, including slow motion and freeze-frame.

*Slow motion* lets you slow things down so you can better analyze fast action, and *freeze-frame* lets you stop the action altogether. For freeze-frame, you find the frame you want, and press the PAUSE (or STILL) button on the VCR. Freeze-frame is available on most models, but slow motion is a feature usually found on high-end models only. For slow motion, you can usually vary the speed from very slow to half the normal speed.

For relatively noise-free pictures—in freeze-frame or slow motion modes—VCRs have one or more special-effects video heads, as discussed in chapter 2, "How VCRs work." In addition, the latest VCRs incorporate digital circuitry to help get rid of the telltale "noise" you see when freezing a frame.

Because of the way special-effects heads work, freeze-frame and slow motion might not be available at all speeds on some decks. Typically, special video effects are not available at the LP speed on a VHS machine or at the BIII speed on some Beta decks.

Many VCRs have the opposite of slow motion, called *fast scan*. With fast scan (also called *cue* or *review*), the action speeds by quickly so you can find a particular spot on the tape without having to go back and forth between play and fast forward or rewind. The fast scan speed varies depending on the machine and the speed that the tape was originally recorded. The speed is typically from three to ten times faster than normal play, depending on the speed at which the tape was originally recorded.

## Stereo ready and stereo adaptable

Being able to record and play back tapes in stereo doesn't mean the VCR can receive television broadcasts in stereo. That requires a deck that's either stereo ready or stereo adaptable.

*Stereo ready* means that the VCR has an *MTS* (multichannel TV sound) tuner built into it, so that without extra circuitry, the deck can receive and process programs broadcast to you in stereo.

*Stereo adaptable* means that the VCR can be connected to a decoder that includes the circuitry necessary to receive and process the stereo sound. Without the box, the TV processes monaural sound only. A VCR that is stereo adaptable has an *MPX*, or multiplex, jack on it for connection to the decoder. This type VCR is no longer made, but you might own an older model with this feature.

## Audio and video dubbing

When you place a VCR in the record mode, it tapes both the sound and picture at the same time. Some top-of-the-line models let you record only audio or video, leaving existing sound or picture signals intact.

Assume you've videotaped your collection of Super 8mm home movies. Unless they're sound movies, you won't have a soundtrack to accompany the picture, but that doesn't mean you have to sit through silent movies whenever you watch the tape. An audio dub facility allows you to record sound after the picture has been placed on the tape, as shown in Fig. 1-15.

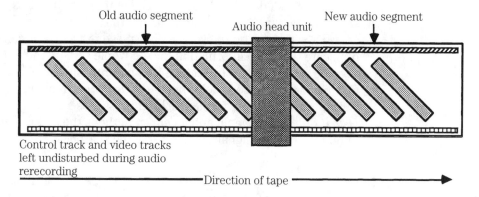

Old audio segment

Audio head unit

New audio segment

Control track and video tracks
left undisturbed during audio
rerecording

Direction of tape

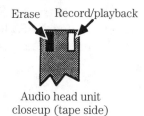

Erase    Record/playback

Audio head unit
closeup (tape side)

■ **1-15** *During audio dubbing, only the audio track is erased and re-recorded. The control and video tracks are unaffected.*

If the VCR is an older linear stereo model, you can usually record the left or right channels independently (see the discussion of audio heads in this chapter). This independence lets you add sound during one recording session (on the right channel, for example) and narration during another recording session (in this case, the left channel). Video dubbing records new images over old without harming the audio portion.

## HQ

Many of the latest VHS VCRs use circuitry to enhance the quality artificially. The circuits are collectively known as *HQ*, which stands for high quality. Here is a summary of the HQ circuits and what they do.

☐ *White clip level* enhancement. All VCRs cut off scenes that exceed a certain brightness. The cutoff point is known as the white clip level. In HQ VCRs, the white clip is extended by another 20 percent, thus increasing the apparent sharpness of details and outlines.

☐ *Brightness noise* reduction. Brightness noise looks like grains of sand sprinkled over the picture. The more noise, the less sharp the picture. Noise is random, and the VCR never knows where it will occur on the screen. The noise reduction circuits duplicate the video information for the whole screen, then average the result, so the picture is apparently sharper.

☐ *Color noise* reduction. Color noise shows up as streaking and is reduced in a similar fashion as brightness noise.

☐ *Detail enhancer*. Detail enhancer strengthens the amplitude of high-frequency signals, which carry the sharpness information.

Not all makers of VHS VCRs have adopted HQ, and not all make use of every one of the four separate circuits described in the preceding list. Often, only two or three of the circuits will be used, such as the brightness noise reducer and detail enhancer or the white clip. A few VCRs let you vary the amount of noise reduction or detail enhancement or turn the HQ circuits off altogether.

## Super VHS

Super VHS is a subformat of VHS in which the picture resolution is increased by separating the brightness and color components and by widening the range of the brightness signal. The unusual signal scheme of Super VHS makes tapes recorded on a Super VHS impossible to watch on a regular VHS deck (exception: some models of regular VHS VCRs can play a Super VHS tape, but cannot record one). However, all Super VHS machines can play standard VHS tapes, and you can manually switch the Super deck to record in normal VHS format. This capability assures you of being able to watch prerecorded movies and share tapes with friends and relatives who do not own Super VHS. There's more about this interesting subformat in the next chapter.

## Super Beta

In 1984, Sony introduced a new signal-processing scheme that squeezed out extra picture quality from Beta decks. The new Super Beta VCRs can deliver a signal up to 20% sharper than standard

**18**

Beta models. The actual increase in picture quality you enjoy also is determined by the quality of tape you're using, the source program, and other variables. If you have a low-performance TV, for example, you won't likely see much of an improvement at all. Tapes recorded on a Super Beta deck can be played back on a standard deck, but without the improvement in picture quality. Likewise, a Super Beta deck can't deliver higher-quality pictures when playing back a tape recorded on a conventional Beta machine.

## Index

The VHS index system is a method using the control track of VHS video recorders to enable you to find a certain segment on the tape quickly. The index system found mostly on mid- to upper-level decks provides two types of coding: *index* and *numeric address*. With index coding, you identify the start or some other portion of a program with "markers." Indexed points are sensed during playback and fast scan (forward or reverse). Depending on the model of VCR, the index codes can be added during or after initial program recording.

The VHS address search system allows you to place a multiple-digit code onto the tape. This address code can be written either manually (by you) or automatically (by the VCR) during recording or playback. You might, for example, use the address code system to number the programs as they appear on a tape, to memorize the play or record time, or even to jot down the time and date of the recording.

## Digital-effects VCRs

Digital-effects VCRs use computer circuitry to produce blemish-free special effects during playback. With all VCRs, pausing the tape during playback yields *noise bars* or lines at the top and bottom of the screen.

Digital circuits freeze an instant of video by capturing the image into computer memory, then relaying the image stored in memory to the TV set. The tape itself can keep moving within the VCR; the image on the screen is frozen in silicon and doesn't change until you hit the FREEZE or PAUSE button again. You can find more details on digital-effects VCRs in the next chapter.

## Jog/shuttle control

Top-end VCRs offer numerous enhanced playback controls, which are particularly handy for video editing. One of the most useful playback controls is the jog/shuttle knob. This knob, placed either on the front of the deck or on the infrared remote control (or both), lets you dial in the exact playback speed and direction.

Twist the knob forward, and the VCR advances into normal playback mode. Twist it some more, and the VCR switches to slow-speed fast forward. You also can use the jog/shuttle knob to advance the tape slowly forward or backward one video frame at a time. This function allows you to easily locate the exact frame you want for editing or cueing.

## On-screen programming and bar-code programming

Programming a VCR to record a program while you're away is probably the toughest task of using a video recorder. The older models, particularly, required you to press a series of buttons in an exact order to program the VCR to record a program at a certain time, on a certain channel, and for a certain duration.

On many VCRs, the programming information is displayed on a special menu screen on the TV (this menu is not recorded on the tape). By having the programming information prominently displayed on the TV screen, you can more easily operate the timer for unattended operation.

For those who really dislike pushing buttons, bar-code programming was all the rage in the mid to late 1980s (particularly on models made by Panasonic or manufactured by Panasonic's parent company, Matsushita). With *bar-code programming*, you moved a wand over a preprinted bar-code swatch. Each swipe of the wand entered a portion of the programmed information.

## Built-in VCR Plus+

The VCR Plus+ programming controller, developed and marketed by Gemstar, has become a common staple in the video den. The stand-alone VCR Plus+ controller looks like a remote control, but it has a special purpose. It lets you record programs simply by punching in a four- to nine-digit number into the VCR Plus+. You get the number from *TV Guide* or other TV listing that publishes the numbers to be used with the VCR Plus+ (these numbers are called *PlusCode* numbers).

Because of the popularity of the VCR Plus+ system, several VCR manufacturers have taken to building the feature into their latest mid- and top-level VCRs. With the VCR Plus+ built in, you can program your VCR to record shows either by using the old-fashioned programming method of specifying the start and stop times and channel, or you can merely look up the PlusCode number for the show you want to record, and punch that into the VCR's remote control.

## Self-cleaning video heads

Over time, the video heads in a VCR can accumulate dirt and grime. Regular cleaning removes the gunk and helps restore the record and playback quality of the VCR. Ordinarily, you clean the video heads using a commercially available cleaning cassette, or for a professional job, you take the cover off the VCR and clean the heads manually with a suitable swab and cleaner.

Some of the latest VCRs come with an automatic video head cleaning system. The mechanism gently wipes off the video heads every time a tape is loaded into the VCR. The self-cleaning system is designed to last as long as the video heads in the VCR—about two to four thousand hours of use.

Both the cleaning cassette and manual cleaning method are recommended for deep-down cleaning. Even if your VCR has a self-cleaning video head feature, you should still periodically clean the heads using one or both other methods, which are described in more detail in chapter 5.

**21**

# How VCRs work

UNDERSTANDING THE WAY VCRs WORK IS AN INTEGRAL PART
of knowing how to repair them when they break. Although routine
maintenance procedures and home repair do not require a scientific
explanation of such things as how the heads in a VCR magnetically
record information on tape, it is helpful to know the basics of VCR
operation.

This chapter details the technical side of the inside workings of
VCRs, with full discussions on the tape loading and threading
schemes used by Beta and VHS decks, how the tape is transported
within the machine, how the number of video heads determines
picture quality at all playback speeds, and more.

In many ways, VCRs are like television sets without the picture
tube. To understand how VCRs play back and record picture and
sound, you need to know how a television works. Hence, this chapter contains details on the science of television—how electronic
images are beamed through the air and received and processed by
your TV and VCR. It's best to understand basic TV first; then grad-
uate to the operation of VCRs.

## Television history 101

Though it seems as if television has been around for a long time,
it's a relatively new science, younger than rocketry, internal medicine, and nuclear physics. In fact, some of the people that helped
develop the first commercial TV sets and erect the first TV broadcast antennas are still living.

The first electronic transmission of a picture was believed to be
made by a Scotsman, John Logie Baird, in the cold month of Feb-
ruary 1924. His subject was a Maltese cross, its image transmitted
through the air by the magic of television (also called "Televisor"
or "Radiovision" in those days) the entire distance of 10 feet.

To say that Baird's contraption was crude is an understatement. His Televisor was made from a cardboard scanning disk, some darning needles, a few discarded electric motors, piano wire, glue, and other assorted odds and ends. The picture reproduced by the original Baird Televisor was extremely difficult to see—a shadow, at best.

Until about 1928, other amateur radiovision enthusiasts toyed with Baird's basic design, whittling long hours in the basement transmitting Maltese crosses, model airplanes, flags, and anything else that would stay still long enough under the intense light required to produce an image. (As an interesting aside, Baird's lighting for his 1924 Maltese cross transmission required 2000 V (volts) produced by a roomful of batteries. So much heat was generated by the lighting equipment that Baird eventually burned down his laboratory.)

Baird's electromechanical approach to television led the way to future developments of transmitting and receiving pictures. The nature of the Baird Televisor, however, limited the clarity and stability of images. Most of the sets made and sold in those days required the viewer to peer through a glass lens to watch the screen, which was seldom more than $7 \times 10$ inches. What's more, the majority of screens had an annoying orange glow that often marred reception and irritated the eyes.

In the early 1930s, Vladimir Zworykin developed a device known as the *iconoscope* camera. About the same time, Philo T. Farnsworth was putting the finishing touches on the image dissector tube, a gizmo that proved to be the forerunner to the modern *CRT* (cathode-ray tube)—the everyday picture tube. These two devices paved the way to the TV sets you know and cherish today.

The first commercially available modern-day cathode-ray tube televisions were available in about 1936. Tens of thousands of these sets were sold throughout the United States and Great Britain, even though there were no regular television broadcasts until 1939, when RCA started what was to become the first American television network, NBC. Incidentally, the first true network transmission was in early 1940, between NBC's sister stations WNBT in New York City (now WNBC-TV) and WRGB in Schenectady.

World War II greatly hampered the development of television, and between 1941 and 1945, no television sets were commercially produced (engineers were too busy perfecting the radar, which, interestingly enough, contributed significantly to the development of conventional TV). But after the war, the television industry boomed. Television sets were selling like hotcakes, even though they cost an average of $650 (based on the average wage earnings for that era, that's equivalent to about $4000 today).

Progress took a giant step in 1948 and 1949 when the four American networks, NBC, CBS, ABC, and Dumont, introduced quality, class-act programming, which at the time included "Kraft Television Theatre," "Howdy Doody," and "The Texaco Star Theatre" with Milton Berle. These famous stars of the stage and radio made people want to own a television set.

Since the late 1940s, television technology has continued to improve and mature. Color came on December 17, 1953, when the FCC approved RCA's all-electronic system, thus ending a bitter, four-year bout between CBS and RCA over color transmission standards. Television images beamed via space satellite caught the public's fancy in July of 1962 when Telstar 1 relayed images of AT&T chairman Frederick R. Kappell from the U.S. to Great Britain. Pay TV came and went several times from the 1950s through the 1970s; modern-day professional commercial videotape machines were demonstrated in 1956 by Ampex; and home video recorders appeared on retail shelves by early 1976.

## What is television?

In an old "Twilight Zone" episode, a gun-toting desperado from the nineteenth century is accidentally time-ported to present day. While walking through the busy streets of New York, he becomes unsettled by cars and bothered by the tall skyscrapers. But he totally falls apart when he sees a television set. There, displayed on the TV, is a classic Western: The good guy tells the bad guy to lay down his gun—or else. What does the desperado do? Why, he draws his gun and blasts the TV to bits. Though the picture displayed on the TV was only in black and white, it looked like the real thing to the OK Corral reject. He saw a sharp, stable picture, both bright and vivid.

What the desperado didn't know is that what he thought was a real picture was actually a beam of light sweeping very rapidly, line by line, over the face of a glass tube. The electron beam in a TV tube moves so rapidly and with such precision that the eye doesn't detect it, but fuses it into a steady, complete picture (the *persistence of vision* that Edison relied on when he invented the motion-picture camera and projector also works with television images). If you were to slow down the beam, however, you'd see it start from the upper left corner, sweep over the face of the tube from left to right and top to bottom, as depicted in Fig. 2-1. This process repeats over and over again.

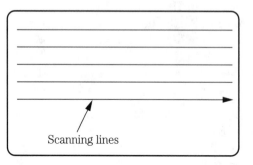

■ **2-1** *The electron beam of a TV tube scanning the face of the screen.*

Scanning lines

## Breakdown of TV signals

The TV tube in the television set is controlled by the incoming TV signal. This signal, after it is tuned to a specific channel and the audio is separated and sent to the audio amplifier, consists of many different subsignals placed on a band that is 4.2 MHz wide. This band is shown in Fig. 2-2. The component TV signals are:

☐ Picture brightness information, called luminance
☐ Color subcarrier signal, called chrominance
☐ 60 Hz vertical sync signal
☐ 15.7 kHz horizontal sync signal

Note that when the signal is transmitted through the airwaves (or sent through a cable), the picture is *AM* (amplitude modulated) and the sound is *FM* (frequency modulated). The audio carrier is placed at 4.5 MHz, which is just outside the luminance envelope.

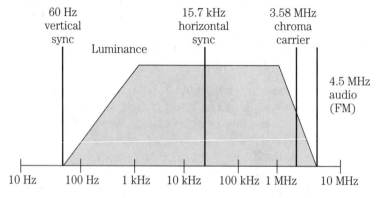

■ **2-2** *A logarithmic view of the television broadcast spectrum.*

The luminance (Y) signal includes the bulk of the television transmission. The image information conveyed in the TV picture is vast, so it requires a large bandwidth, essentially from 100 Hz to 4.2 MHz. Note that the frequency of the information conveyed in the luminance signal determines the detail of the picture. A low-frequency signal creates a large, indistinguishable blur on the screen, and a high-frequency signal creates a well-defined, pinpoint shape.

The color (C) subcarrier signal, called *chroma* (also *chrominance* or *burst*) is present only during color transmission. The science of color television transmission is complex and therefore beyond the scope of this book. But it is sufficient to say that the chroma subcarrier contains the information for both the color and color intensity for every instant of the transmission. The chroma subcarrier is phase dependent; when the phase of the signal changes, the color changes. The instantaneous amplitude of the chroma signal determines the intensity or saturation of the color.

Keep in mind that there is more to color TV than this, and if you are interested in learning more, refer to appendix A, "Sources," for a list of books that discuss the topic in more detail.

The 60 Hz vertical synchronization is used by the television picture tube to direct the beam from the bottom of the screen back up to the top. This process is repeated 60 times per second, so the vertical sync rate is 60 Hz.

The 15.7 kHz horizontal synchronization is used by the television picture tube to direct the beam from left to right over the inside face of the tube. This process is repeated 15,734 times per second, so the exact horizontal sync rate is 15.734 kHz. For convenience, the horizontal sync rate is normally expressed as 15.7 kHz.

Note that all of these signals do not occur at the same time. In fact, they are precisely timed so the TV set can adequately react to them, and in the proper order. Figure 2-3, as shown on the next page shows a brief moment of television transmission. It illustrates the stream of signal transmitted to the television set and how it is broken down into the components discussed.

## Inside the picture tube

If you dissect a picture tube, you can more fully understand the role that these signals play (for the time being, ignore the chroma signal). An electron gun is at the base of the TV tube. It emits a steady stream of electrons that strike the inside face of the tube, which is coated with a phosphor. When the electrons impinge on

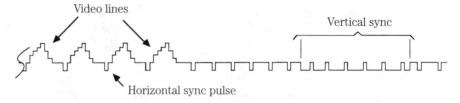

**2-3** *A small portion of the steady stream of signals sent during TV transmission. Note the horizontal sync pulses between each scanned line of video.*

the phosphor, the phosphor emits light, causing a glow on the outside of the front of the tube.

The luminance signal controls the intensity of the electron beam. Because the amplitude of the luminance signal is always changing, the intensity of the beam is always changing. The horizontal and vertical sync signals operate magnets that are located around the neck of the tube. The vertical deflection magnets are controlled by the vertical sync signal, and the horizontal deflection magnets are controlled by the horizontal sync signal.

## Of lines, fields, and frames

One horizontal line of picture information is defined as the electron beam tracing the TV screen from left to right one time. At the end of the line, the TV set receives a horizontal sync pulse, so the beam shuttles back to the left side. Circuits inside the TV set start the beam a little lower in the screen each successive time so it doesn't trace the same spot over and over again.

The process repeats 262½ times before the electron beam reaches the bottom center of the screen. All of the lines traced so far make up one field. Once the field has been traced, the vertical sync pulse pulls the electron beam back up to the top of the screen (now starting at the center). The beam now sweeps over the screen another 262½ times, but this time, the new traces are between those previously made (see Fig. 2-4). When the beam reaches the bottom right corner of the tube, the vertical sync pulses again pull it back to the top and the process starts over again.

The two fields together make a frame. There are 60 fields per second and 30 frames per second. The pair of 262½ line fields comprise a total number of 525 lines per frame. The lines of the two fields are melded together with what's known as 2:1 interlace. The first field of the frame traces the odd-numbered lines (1, 3, 5, etc.), and the second field of the frame traces the even-numbered

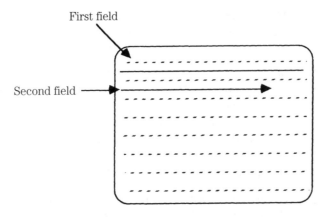

First field

Second field

**■ 2-4** *The electron beam sweeps between the previously scanned lines to generate the second field of the video frame.*

lines (2, 4, 6, etc.). The 2:1 interlace allows for less complex circuitry, yet the *refresh rate* of the screen (the time between each new successive frame) is high enough that flicker is not visible.

### Adding color

In a black-and-white set, there is one electron gun. In color sets, there are three electron guns, or at least one gun that produces three separate electron beams. The three beams are deflected together by the horizontal and vertical sync magnets and strike the face of the tube in a triangular pattern. Each beam hits a differently colored phosphor, either red, green, or blue. The intensity of the three colors at any one time determines the final color as seen on the outside of the tube.

For example, if the green phosphors are lit but the others are not, you see an all-green picture. But if the red and blue phosphors are lit, but the green ones are not, you see purple. All the colors in the rainbow, and several million others, are derived from the three primary television colors.

## Baseband and RF signals

When TV picture information is combined with chroma and sync signals, it is called *composite video*. Composite video is the lowest level of TV signal that you will encounter in home video setups. In contrast, professional broadcast systems use equipment where the luminance signal is often separate from the sync and chroma signals. (Exception: Super VHS VCRs have separate luminance and chrominance inputs and outputs. The Y-C output of the VCR is intended to mate with the Y-C terminals on a suitable television set. There is more about Super VHS and how it works in this chapter.)

When composite video is not modulated on a carrier signal by any means, it is referred to as *baseband video*. The same is true of the audio portion of the program. Unmodulated audio is termed *baseband audio*. As you probably already know, baseband video and audio signals cannot be sent through the airwaves or down a cable; they must be carried by a modulated signal, one that combines both video and audio and is received over a specific TV channel.

See how this works with the typical neighborhood channel 4 TV station. The signals are transmitted from the TV station antenna as *RF* (radio frequency) signals. Specifically, the channel 4 broadcast is AM and FM modulated in the 66 to 72 MHz range. The picture carrier is centered at 67.25 MHz, the color subcarrier is at 70.83 MHz, and the sound carrier is centered at 71.72 MHz.

To watch the station, you dial to channel 4 on your TV or VCR. The tuner in the TV or VCR then receives the signals sent at the frequencies listed in this chapter. After being processed by intermediate stages, the modulation carriers are stripped from the transmission and baseband audio and video result.

All tabletop VCRs have an RF output with an output frequency of either channel 3 or 4. You connect the RF output of the VCR to the VHF terminals of your TV, then tune the TV to channel 3 or 4.

If you have a VCR, you've probably noticed that it has both an RF output and separate video and audio outputs. The same sound and picture information is available at both outputs, it's just in different forms. The RF output is modulated to TV channels 3 or 4. The video and audio outputs carry baseband composite video and audio. You use one or the other output, depending on your system. Because of the extra processing stages inherent in RF reception and transmission, use baseband audio and video whenever possible.

## Basic VCR block diagram

The basic operation of all VCRs is the same. Figure 2-5 shows the major sections of a VCR, including the record/playback heads, transport mechanism, safety devices, and so forth. Note that not all VCRs have every block. For example, the control circuit is found only on models built after 1983 or 1984, and camcorders lack a timer and tuner.

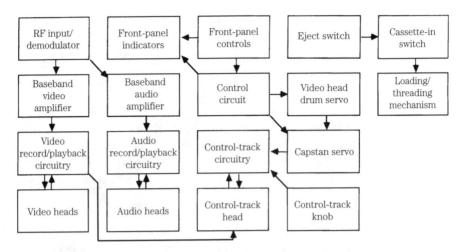

**2-5** *A basic block diagram of the typical VCR. The central control unit in some decks is a microprocessor or one-chip computer.*

## Cassette loading

Home VCRs use a tape enclosed in a protective cartridge, also known as the *shell* or *cassette*. There are two types of cassette-loading schemes: front and top.

### Front loading

Front-loading VCRs are the mainstay, even though the loading mechanism is more complex than top-loading models (see the following section). Front loading allows you to fit the VCR in a tight slot in the front of the video system and not worry about obstructing the lift mechanism. With a top-loading VCR, you must leave the top free to access the cassette lift mechanism.

To load a tape, insert it (hubs down, door to the front) into the loading slot (also called the *dock*). When the cassette has been inserted about midway, it triggers a leaf switch. This tells the VCR that a tape is inserted. The VCR then latches onto the tape and pulls it all the way into the cassette lift mechanism (also called the *elevator* or *basket*). After the cartridge is in the VCR, the cassette lift mechanism drops down. When loaded, the reels of the cassette rest over the supply and take-up spindles inside the VCR. The tape can now be threaded and played.

To unload the tape, press the EJECT button (on some VCRs, you press the STOP button twice to eject the tape). The cassette lift mechanism lifts the cassette up and pushes it out of the slot where you can retrieve it. On most VCRs, hitting EJECT removes the tape even if the deck is playing. A mishap won't occur, however, because the VCR first stops the tape and unthreads it.

As with the top-loading units, front-load Beta decks automatically thread the tape immediately after loading. The tape remains threaded until you eject the cassette. Also, a number of front-loading VHS decks use a half-loading or partial-loading tape system where the tape is threaded partially around the inside of the deck. This minimizes tape warping, and allows the tape to start playback quicker when changing between modes (such as fast-forward to play).

## Top loading

Top-loading VCRs, where the tape is inserted into a lift mechanism on top of the deck, are no longer made (with one exception—the camcorder). To load the tape into the VCR, you press the EJECT button, which pops up the lift mechanism. You then slide the cassette in so that the reel hubs are on the bottom of the cartridge and the tape cover is in the front. After the tape is inserted, manually press down on the cassette lift mechanism. Loading is then complete. When loaded, the tape reels of the cassette rest over the supply and take-up spindles inside the VCR.

To unload the cassette, stop the VCR (if it isn't already), and press the EJECT button. The cassette lift mechanism pops back up and you remove the cassette.

In top-loading Beta VCRs, the tape is automatically threaded around the tape transport mechanism and is unthreaded when you press the EJECT button. Because the tape is always threaded in a Beta deck, tape stress can cause buckling, which can lead to jams. That's why you should never leave a tape in a Beta deck when the machine is not actually in use.

## Tape threading

The transport mechanism of a VCR is considerably more complex than the transport of a reel-to-reel or cassette audio tape deck. In the older *VTRs* (videotape recorders), you were required to thread the tape manually around the heads and through the capstan and pinchroller. The cassette design of VHS, 8mm, and Beta decks relieves you from this requirement, so your hands never have to touch the tape.

VHS and Beta decks use dramatically different threading mechanisms, yet the end result is nearly identical. Take a closer look at the threading characteristics of each format.

## VHS threading

Since their introduction, all VHS decks have had the same threading procedure. When the tape is loaded into position, two tape-threading guides, a tape tension lever, and a capstan protrude into the cassette area, as shown in Fig. 2-6. Threading is accomplished by the two tape-threading guides. These guides, which are mounted on either side of the video-head drum, pull the tape out of the cartridge and wrap it around the drum. The erase, audio, and control-track heads are positioned so the tape is pressed firmly against them once threading is complete.

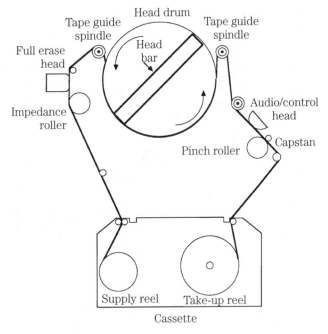

**■ 2-6** *A simplified view of VHS threading.*

During threading, the pinchroller comes into contact with the tape and capstan and acts as a pressure roller for tape transport (see the following section). The tape tension lever, mounted on the left side of the transport area, swings to the left. The lever helps maintain even back pressure on the tape and prevents the tape from spilling out into the transport area.

The VHS threading path layout takes on the shape of the letter *M*, so VHS decks are said to use *M loading*. VHS-C and 8mm decks use similar loading techniques.

## VHS half-loading mechanism

Many of the newer VHS decks, particularly the mid- to upper-end models, incorporate half-loading mechanisms. The feature really ought to be called *half threading* because the cassette is fully loaded into the deck, but the tape is only partially threaded around the transport inside the unit.

Figure 2-7 shows a VHS transport mechanism with full and half loading. Note that the tape still contacts the video-head drum and is in contact with the audio/control head and capstan. Half loading retains such advanced features as VHS index or real-time counters. In the latter, the tape counter shows actual elapsed seconds, minutes, and hours, not just an arbitrary number. The deck automatically switches between full and half loading, depending on the current action; see Table 2-1.

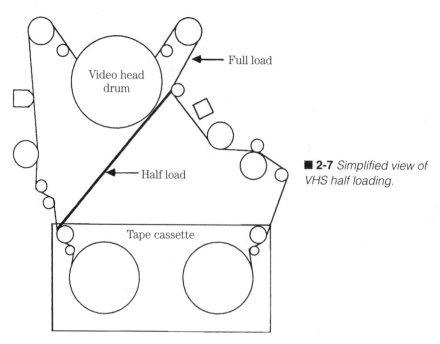

■ **2-7** *Simplified view of VHS half loading.*

When the cassette is inserted into the VCR, the tape is immediately half loaded and remains that way until you press PLAY or RECORD. The tape will return to half-load threading when you press STOP (pressing FAST FORWARD or REWIND during play activates the fast-search mode; the tape remains fully loaded, and the high-speed picture is displayed on the TV).

| Action | Half loading | Full loading |
|---|---|---|
| Play | | X |
| Record | | X |
| Search | | X |
| Still | | X |
| Frame advance | | X |
| Fast forward | X | |
| Rewind | X | |
| Stop | X | |

## Beta threading

Figure 2-8 shows a simplified diagram of the threading mechanism of a modern Beta deck. When the cassette is loaded, the VCR opens the door on the cartridge that protects the tape. That allows the VCR to gain access to the tape and pull it out of the cartridge.

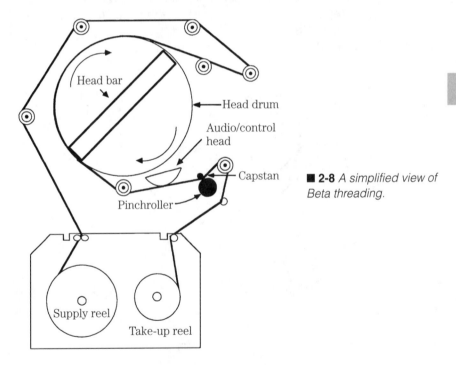

■ **2-8** *A simplified view of Beta threading.*

When the tape is loaded into position (either with the manual cassette lift mechanism in top-loading machines or the automatic method of front loaders), several threading guides and posts protrude inside the cassette area. These include a main threading post and a capstan. The main (or *lead*) threading post is one of several posts mounted on a threading ring.

To initiate threading, the VCR activates a threading motor. This turns the threading ring, thus pulling tape out of the cartridge. The ring revolves around the video-head drum and wraps the tape more than half way around it. Additional tape guide posts on the ring support the tape as it is threaded into position. When the ring finally stops turning, the tape is positioned against the erase, video, audio, and control track heads.

Some decks use additional rollers and guide pins. With a little imagination, you can visualize the threading layout as the letter U. The U-load technique also was used in the ¾-inch videocassette system pioneered by Sony in the early 1970s (that system is still used professionally).

Note that early-model Beta decks used a different threading path layout. The major difference, as shown in Fig. 2-9, is that the tape is wound around the video head in the opposite direction. What is not as apparent is that the older threading system induced excessive tape stress and was prone to the problems of misthreading. The newer Beta threading system poses little stress on the tape and the difficulties of misthreading are greatly reduced.

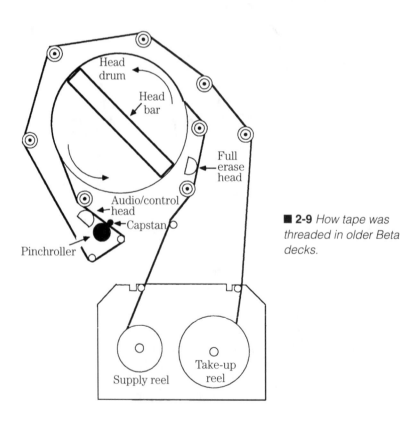

■ 2-9 *How tape was threaded in older Beta decks.*

### Unthreading

The threading mechanism in the VCR works in reverse to un-thread the tape. In Beta decks, the threading ring unwinds while the supply spindle back spins to take up the excess. In VHS decks, the two threading posts retract, and the supply spindle back spins to remove slack.

## Tape transport

VHS and 8mm decks thread the tape only for play and record (plus fast-scan or search modes). The tape is unthreaded for fast forward and rewind. Conversely in Beta VCRs, the tape is always threaded until the EJECT button is pressed.

With the tape threaded, the deck is ready for recording or playback. Several things occur within the tape transport mechanism during recording and playback:

- ☐ The capstan and pinchroller pull the tape through the transport and past the heads.
- ☐ The take-up spindle winds the tape that has passed through the transport onto the take-up reel.
- ☐ The video-head drum spins at 1800 revolutions per minute (30 revolutions per second) to record and play back video.

### Head drum

The video-head drum spins under propulsion from a direct-drive motor (in some early VCRs, the drum was operated by an ac or dc servo-controlled motor). This direct-drive motor contains no brushes, just coils and magnets. The coils in the motor (usually three) are switched on and off by a servo circuit.

Precise timing of the switching is accomplished by a novel feed-back mechanism provided by a set of Hall-effect integrated circuits (ICs). The Hall-effect ICs, which are the electronic equivalent of reed switches, are magnetically sensitive and can detect the presence or absence of a magnetic field. In older Beta and VHS models, coils and magnets are used instead of Hall-effect sensors.

The Hall-effect ICs generate two types of pulses: PG and FG. PG stands for pulse generator; FG stands for frequency generator. The PG signal, typically at 30 Hz, identifies the position of the drum as it spins. The rate of the FG signal, most often 180 Hz, indicates the speed of the video-head drum. As the drum rotates, the Hall-effect sensors relay the position and speed of the drum to the servo.

The servo, which is clocked to a precise standard, increases or decreases the driving signal to the video-head drum motor. The driving signal is pulse-width modulated—often referred to as *PWM*. In the PWM system, the same voltage is always applied to the motor, but the duration of the pulses are shorter or longer, depending on whether the rotation of the drum is to remain the same, speed up, or slow down.

During recording, the servo is governed by a 60 Hz pilot signal. On the old VCRs, that signal was usually derived from the ac line frequency; now the signal is invariably created by the 3.58 MHz color subcarrier quartz crystal (divided down to 60 Hz by a counter or set of flip-flops). With each revolution of the video-head drum, a short tone or pulse is placed on the control track by the control track head. The control-track head records 30 pulses per second. During playback, the servo is maintained by the signal encoded on the control track as well as a clock circuit on the VCR.

## The role of the control-track head

The control-track signal serves three basic functions:

☐ The signal identifies the speed of the recording so the VCR can adjust its playback speed accordingly (recording speed is set by a switch, but playback speed is automatic).
☐ The signal maintains proper playback speed even when the tape stretches or shrinks.
☐ The signal ensures that the video heads trace the video tracks made during recording.

Without the control track, the VCR would not be able to operate. If the control track becomes damaged or the control-track head becomes dirty, proper recording and playback is impossible.

Beta and VHS VCRs are equipped with a tracking-control knob to fine tune the timing between the control-track pulses and scanning of the video head. In normal operation, the control is centered in its detent position. Tracking errors, which appear as lines in the top and/or bottom of the picture, are corrected by adjusting the tracking control to retard or advance the timing between the control-track pulses and the motion of the video heads. The 8mm VCRs lack a control-track knob; pilot signals are recorded on the tape that serve to identify the tracks made by each head during recording.

## Capstan

The capstan is driven via a belt or roller by a precision dc motor, as shown in Fig. 2-10. The speed of the capstan determines the rate at

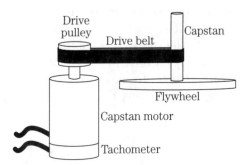

■ **2-10** *The capstan and flywheel are driven by the capstan motor through a drive belt.*

which the tape is pulled through the transport. If the speed varies, both the audio and video are affected. Wow and flutter are induced in the audio when the capstan speed is not properly regulated. The picture might roll or jump, and heavy lines might appear all through the screen. If the speed variations are great, the picture can be muted completely by the VCR circuits, and you'll see nothing.

Small speed imperfections are smoothed out by the capstan flywheel, a heavy metal wheel that uses inertia to keep the capstan speed more or less constant. Precise control of the capstan is maintained via electronic means.

The speed of the capstan is controlled electronically by a two-element process: frequency-related FG tachometer pulses, generated by the capstan, and position-related PG timing pulses, originally generated by the video-head drum. The PG pulses from the video-head drum serve as a reference, and the FG pulses generated by a tachometer built onto the capstan serve as feed-back.

At any given recording or playback speed, the VCR examines the rate of PG pulses coming from the video-head drum (which are always the same) and compares them to the rate of FG pulses coming from the capstan (which change depending on the tape speed).

As with the video-head drum, the VCR changes the speed of the capstan motor not by varying the voltage but by lengthening or shortening the time (or *duty cycle*) the capstan motor receives full voltage.

Note that testing the capstan motor input voltage with a meter yields different voltages, depending on the playback speed. Why? The volt-ohmmeter cannot react to the fast PWM pulses delivered to the motor, so it averages them out. During EP play, when the tape is traveling the slowest, the meter reading might be something like 5 V. The actual voltage to the motor could be approximately 12 V, but the average voltage is much lower because the motor is receiving power for only a short time.

During SP play, however, the motor receives power for a longer time, so the average voltage can double or triple. In fast-scan, the voltage can swing close to the full 12 V. Keep this in mind when testing the capstan in your VCR. You can test overall operation by connecting the leads of the meter to the power terminals, but you have no idea if the motor is receiving the correct duty cycle for that given speed.

## Take-up spindle

During record and playback, the take-up spindle turns to wind the tape onto the take-up reel inside the cartridge. In some VCRs (most notably those made by Mitsubishi), the take-up spindle is driven by a separate motor. The speed of the motor is regulated to keep the tape taut. If the motor stops, or doesn't turn fast enough, the capstan will pull too much tape through the transport and excess tape will accumulate in the VCR, causing a jam.

In most VCRs, the take-up spindle is driven by a roller. In turn, the roller, as shown in Fig. 2-11, is attached to a belt that connects to the capstan motor. This not only shaves a few dollars off the manufacturing price of the VCR, but it helps ensure that the take-up reel travels in proportion to the capstan. The faster the capstan turns, the faster the take-up spindle turns.

As with a reel-to-reel audio deck, the free-running speed of the take-up spindle is greater than the speed of the capstan. If you release the pressure on the capstan exerted by the capstan pinchroller, the take-up spindle will pull the tape through the transport at increased speed. In fact, this is how most decks accomplish fast forwarding. Because the capstan is not limiting the maximum speed of the tape, the tape is wound at the free-running speed of the take-up spindle.

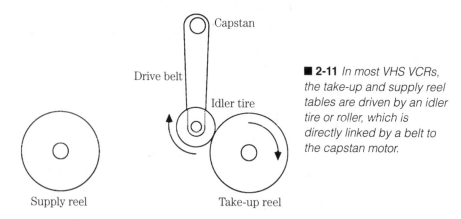

Capstan

Drive belt

Idler tire

Supply reel

Take-up reel

■ **2-11** *In most VHS VCRs, the take-up and supply reel tables are driven by an idler tire or roller, which is directly linked by a belt to the capstan motor.*

Note that the torque of the take-up spindle during normal recording and playback is controlled by a torque limiter. This limiter, usually a pad or brake, prevents the take-up spindle from exerting too much force on the tape, and therefore stretching or breaking it.

### Supply-reel spindle

All VCRs have a supply-reel spindle as well. The spindle is normally unpowered during recording, playback, and fast forward but is powered during reverse play (certain VCRs only), reverse search/fast scan (or *review*), and during rewind. The same belt and roller that drive the take-up spindle also drive the supply spindle. An idler tire swings back and forth, as shown in Fig. 2-12, to deliver power to either one spindle or the other.

Note that the condition of this idler tire is critical to the performance of your VCR. Almost all VCRs use this tire, although a few models incorporate plastic gears (which are prone to stripping and breaking with age). If the tire becomes worn, glazed, or dirty, proper speed of the supply and take-up reels is not maintained. The error in speed invariably leads to tape spillage.

# Recording and playback operations

VCRs use magnetic heads to record and play back video, audio, and control information. Take a few moments to review how magnetic heads operate.

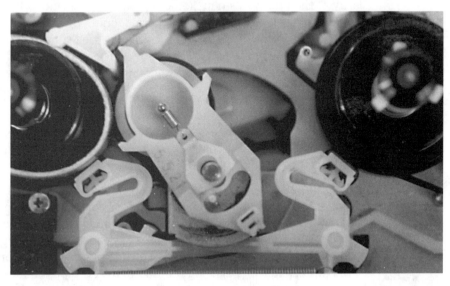

■ **2-12** *Closeup of the idler tire (attached to a pivoting arm) in a VHS deck.*

## The role of magnetic heads

The basic idea behind magnetic head operation is simple. To record, an ac signal is applied to a coil that is wound around a metal core. The core is split down the middle to create an open-air gap, as shown in Fig. 2-13. Without the gap, the head would not operate.

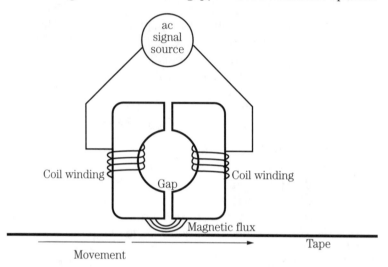

■ **2-13** *Magnetic heads record a signal by passing an alternating current through a coil winding.*

A magnetic *flux* (field) is induced at the gap and changes as the input signal to the head changes. The intensity and magnetic polarity of the flux varies depending on the intensity and polarity of the incoming ac signal. To make a permanent copy of the signal, a tape coated with a magnetically sensitive ferrous substance (usually ferric oxide) passes under the head. The magnetic flux produced by the head is imparted to the tape.

For playback, the tape is passed under the head, but this time no signal is applied to the coil (that would rerecord another signal). The magnetism of the tape causes minute electrical changes in the coil of the head. These changes are amplified and sent to the appropriate circuitry for playback.

Though there is more to magnetic recording and playback to this (such as the addition of a high-frequency bias signal to record faithfully baseband-level audio and control-track signals), the discussion is sufficient for your purposes. The idea is that once the tape is magnetized by the heads, the magnetism remains (more or less) until something else is recorded onto the tape, obliterating the previous recording.

## Signal conversion for recording

The TV spectrum covers a considerable frequency range of 4.2 MHz (that's just for the picture; the range is 4.5 MHz with the audio signal added). Consumer-level magnetic tape does not have the frequency response to record information accurately over such a wide range, so to place audio and video onto a VHS, Beta, or 8mm cassette, the signals must be modified. The modification is extensive and varies slightly depending on the format. The VHS format is discussed here; then the differences found in Beta and 8mm VCRs are explained.

## Luminance and chrominance conversion for recording

After the RF signal is received by the VCR tuner, it is demodulated (stripped of the modulation carriers), the video portion is sent to the video circuits, and the audio portion is sent to the audio circuits. With few exceptions (as with Beta hi-fi decks, detailed subsequently), the audio and video never meet again until the taped program is played back and viewed on your television set.

The composite video then passes through two filters, as shown in Fig. 2-14. One is a lowpass filter, the other a bandpass filter. The lowpass filter passes just the luminance component of the video signal and rejects the chrominance component, which is situated at approximately 3.58 MHz. Follow the luminance signal for now.

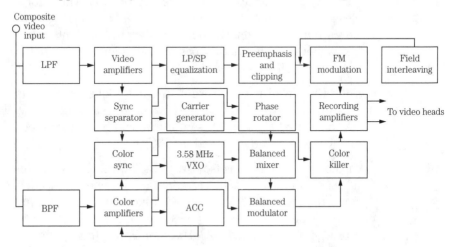

■ **2-14** *The VHS recording circuit.*

Once separated by the filter, the luminance component (or channel) is amplified. The signal is equalized depending on the recording speed (SP, LP, or EP). Equalization is important because it prepares the signal for optimum recording at the desired record/playback speed. Preemphasis and clipping are added to increase high-frequency response and signal-to-noise ratio and to prevent the signal from exceeding defined upper and lower limits.

The composite video signal is then FM modulated with a carrier signal so the total deviation of the video signal spans a 1 MHz range between 3.4 and 4.4 MHz. The FM signal, along with the lower luminance channel sideband, is amplified and applied to the video heads.

Of course, this isn't the whole story. After going through the low-pass filter and video amplifier, the horizontal synchronization signals are detected and separated by a sync separator. One side of the sync separator is applied to a color sync generator (discussed in this chapter), and the other goes to a carrier generator and phase rotator.

The carrier generator creates a 629 kHz *cw* (continuous-wave) signal that feeds the phase rotator. The phase rotator is used to rotate the vector or phase of the color subcarrier signal 90° for each video line. The rotation helps prevent crosstalk between adjacent lines of video.

The chrominance component is separated by the bandpass filter and applied to a color amplifier. The output of the amplifier is controlled by an automatic color corrector, or *ACC*. The color section uses a gating signal created by the horizontal sync rate to separate the chrominance signal. When the signal is present, it turns the color killer off. That action passes the color-burst signal created within the VCR to the heads.

When the signal is not present, the color-killer circuit turns on. That prevents the color-burst signal from reaching the heads. If the color-burst signal is present during a monochrome recording, an objectionable effect called *moiré,* or *color-crawl,* can occur.

The color-sync circuits also drive a 3.58 MHz *VXO* (variable crystal oscillator). The oscillator generates a color-burst-locked cw signal that is fed into a balanced mixer. The other input of the mixer is the output of the phase rotator, described earlier. The output of the mixer is a 4.2 MHz color signal. It is applied to a balanced modulator, which is also driven by the 3.58 MHz signal from the color amplifier. The product is a difference heterodyne that results in a 629 kHz chroma-signal sideband. This 629 kHz signal is coupled with the luminance signal and applied to the video heads.

The technique of transposing the color to a frequency below the luminance signal (or channel) is often referred to as *color-under* (the process is technically known as heterodyning). The waveform of the luminance and chrominance signals, as recorded by the VCR, is shown in Fig. 2-15. Note how it differs from Fig. 2-2, which shows the broadcast television picture spectrum.

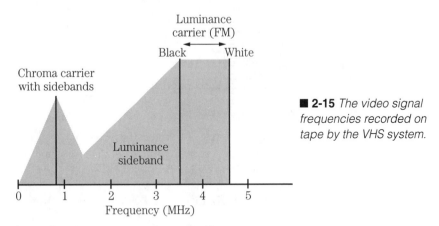

■ **2-15** *The video signal frequencies recorded on tape by the VHS system.*

## Luminance and chrominance conversion for playback

The process is almost reversed for playing back the picture. Refer to Fig. 2-16 on the next page as you read the text. The previously recorded signal is detected on the tape and amplified by the head preamplifiers. The luminance channel is then processed by a dropout compensator that duplicates the information from previous lines to fill in blank areas of tape (caused by a loss of oxide coating on the tape). The *DOC* (dropout compensator), which is standard equipment on all VCRs, limits the appearance of black and white specks on the screen.

Once compensated, the signal is FM demodulated and is returned to its original baseband component. Deemphasis removes slight signal alterations that were made depending on the speed of recording. The output is then applied to a Y-C matrix, which combines the luminance and chrominance channels together.

The 629 kHz color-carrier rotating sidebands are sent to an RF amplifier and balanced modulator. The rotation is removed, thanks to sync separator circuits driven by the luminance component. The color signal is then amplified by a chroma amplifier that drives the color killer, *APC* (automatic phase control), and *ACC* (automatic color control) sections. The output of the color-killer section when the color-burst signal is present is a corrected (nonrotating) color

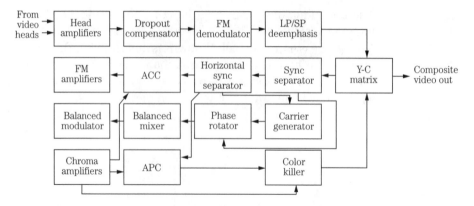

**■ 2-16** *The VHS playback circuit.*

carrier. The output of the Y-C matrix is baseband video and is routed to the VIDEO OUT terminals of the VCR or sent to the RF modulator section of the VCR.

## Audio signal recording and playback

After the baseband audio and composite video signal are separated, the audio portion is sent to the audio recording circuits. These circuits work identically to those found in a reel-to-reel tape recorder. The signal is mixed with a high-frequency bias current (which improves linear response) and is recorded on tape as a baseband signal. The linear audio track is along the top of the tape. In linear stereo (in VHS decks and European Beta decks) there are two discrete tracks, as shown in Fig. 2-17.

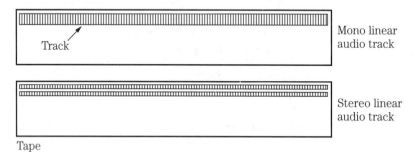

Mono linear
audio track

Stereo linear
audio track

Tape

**■ 2-17** *Monophonic and stereophonic linear sound tracks.*

For playback, the heads pick up the audio track and amplify it. If noise reduction is used, such as Dolby, the audio is processed through the noise-reduction circuitry. The amplified audio is applied to the AUDIO OUT terminals of the VCR and also is sent to the RF section.

## Control-track signal recording and playback

The vertical sync signal is not recorded with the rest of the video because VCRs don't require vertical sync the same way as TV sets. The equivalent of vertical sync is recorded on the control-track heads. As discussed previously, the 30 Hz signal is used to properly time the video heads in relation to the tracks on the tape.

## Beta VCRs

The process of recording and playing back video, audio, and control-track information is similar with Beta models. Figure 2-18 shows a block diagram of the Beta recording function, and Fig. 2-19 shows a block diagram of the Beta playback function.

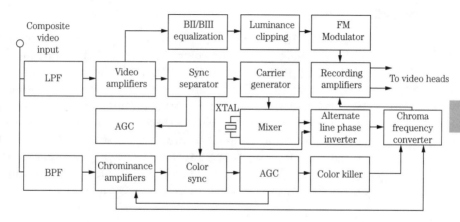

**2-18** *The Beta recording circuit.*

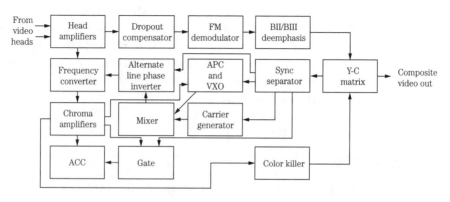

**2-19** *The Beta playback circuit.*

The biggest differences in the recording process are the FM deviation frequency for the luminance channel, the chrominance subcarrier frequency, and an alternate-line phase-inverter circuit. In Beta decks, the luminance channel is frequency modulated to between 3.5 and 4.8 MHz—a 1.3 MHz range (this wider bandwidth is what has always given Beta the edge over VHS). The color channel is down-converted to 688 kHz. On each line of video, the phase of the color-burst signal is inverted 180 degrees. As with the VHS system, the phase inversion reduces crosstalk between adjacent video lines.

The system is nearly reversed for playback. As with VHS, the Beta luminance signal is processed through a dropout compensator. The alternate-line color subcarrier signal is corrected, up-converted to 3.58 MHz, and mixed with the luminance component in the Y-C matrix circuit.

## 8mm VCRs

VCRs in the 8mm camp are, in many ways, scaled-down versions of VHS machines. The same color-under recording technique used in VHS and Beta is used in 8mm except that the chrominance carrier is centered at 743.4 kHz and the luminance carrier is centered at 4.98 MHz. An *ATF* (automatic track finding) signal, recorded 14 dB (decibels) below the level of the chrominance carrier, eliminates the need for manual tracking control adjustment. The ATF system uses four pilot tones to mark successive tracks. During playback, the ATF circuitry identifies the tones and continually adjusts the tracking so the tones are steady and equal.

# Video head geometries

VCRs use a minimum of two video heads to record and play back picture information. Each head records a single TV field in one pass, so two adjacent tracks on the tape make up an entire video frame. In both VHS and Beta systems, the heads are mounted 180 degrees apart, and only one head touches the tape at any time (in VHS-C camcorders, the tape wraps around the head 270 degrees and two heads record on the tape at the same time. Some other camcorders have other unusual geometries, but home VCRs are designed about the same).

## Head gap size

The size of the gap found in all magnetic recording heads varies from about 28 to 90 microns wide, depending on the application. The video heads in typical VHS decks have a head gap of only 38 microns—too small for you to see without a microscope or high-powered magnifier. The gap in conventional Beta video heads is even smaller, 30 microns, and in 8mm decks, the gap is 20.5 microns.

VCRs use head gaps of various sizes to increase performance at slower record and playback speeds and to accomplish noise-free special effects. Wider heads record a stronger signal and are desired over narrow head gaps. But as you can soon see, a problem occurs when using wide heads to record and play back video at the slower LP, EP, or BIII speeds.

Figure 2-20 shows the geometry of the video tracks laid down by the video heads with a conventional VHS VCR at SP, LP, and EP speeds. VHS design parameters allow for 58 microns between each video track at SP speed. The spacing decreases proportionately at the slower speeds—29 microns and 19 microns for LP and EP speeds, respectively. At the fast speed, there are guard bands as wide as 28 microns between the tracks. As the tape slows down, the tracks get closer together and they overlap by as much as 11 microns.

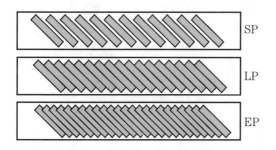

SP

LP

EP

■ **2-20** *Track spacing at the three VHS speeds of SP, LP, and EP (SLP).*

Obviously, if the deck is equipped with wide heads, playback and recording will be satisfactory at the SP speed because of the guard bands between the tracks. But operation will be severely limited at slower speeds because the tracks will almost stack on top of one another during recording.

To provide optimum performance at both fast and slow speeds, many VCRs use four heads instead of just two. The gap is wider on one side of the heads. The wide set is designed for SP speed, and the narrow set is for LP and EP speed. The actual size of the head gap varies depending on the make and model. Figure 2-21 on the next page shows the head geometries for popular two- and four-head VHS designs from JVC and Matsushita. As shown on the next page. Note than the head gap is not always the same for opposite heads in the same set. This provides the deck with better performance during playback and with certain special video effects such as still-frame and slow motion.

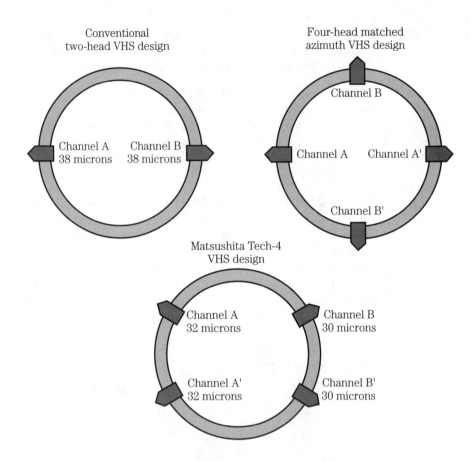

**■ 2-21** *A sampling of head geometries of typical VHS machines.*

### When heads become dirty

When audio heads acquire an excessive build-up of tape oxide and other contaminants, the sound output decreases and might be fuzzy and indistinct. But if you haven't trained yourself to listen for the effects of dirty audio heads, you might not be aware of it.

There is no such blissful ignorance with dirty video heads. When one or more of the video heads in the VCR become dirty or clogged with foreign matter, the picture becomes washed out. If the accumulation of junk is excessive, the video heads won't be able to pick up anything on the tape, and you will see a completely snowy picture.

Many people mistakenly assume that foreign material gets caught in the gap of the heads. The gap itself is sealed with epoxy or other hard, nonferrous substance so nothing can get inside. Rather, the contamination is completely on the surface. But the extremely smooth surface of the video head means that particles of tape oxide, dust, cigarette smoke, and so forth, can adhere like glue. All VCRs require that their heads be cleaned regularly. Other chapters address the procedure for cleaning heads and the proper care that must be exercised.

## Opposing azimuth

Even at the slow LP and EP speeds with a narrow head, the video tracks still overlap. The effects of this inevitable overlap are precluded by canting the video heads at opposing angles. In VHS systems, one video head is tilted +6° to the center and the other is tilted –6° to center, as shown in Fig. 2-22. The 12° difference in so-called azimuth reduces or eliminates the crosstalk that would otherwise occur if the heads were on the same plane. In Beta decks, the azimuth is +7 and –7°, and in 8mm the heads are angled ±10°.

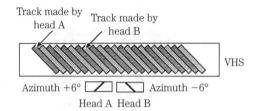

**■ 2-22** *Alternating video tracks are recorded at different angles to minimize cross-channel interference.*

51

## Video heads and noise bars

During playback, the video heads trace the same path they took when recording the signal. If the speed of the tape increases, however, the heads can no longer trace over just one track at a time because the angle of the heads has changed in relation to the tracks recorded on the tape. The result: the heads cross into adjacent tracks. In one pass, the video head might scan a portion of two, three, or even more tracks. The effect you see on the screen is a picture composed of stripes from each successive video track.

The stripes are separated by noise bars, as shown in Fig. 2-23 on the next page. These bars can never be totally eliminated in consumer VCR equipment, but they are reduced with wide heads or when recording with regular heads at slower speeds. With extra wide heads at SP speed, for example, there is little guard band space between the tracks. But the guard bands are wider when recording at SP speed with narrow heads. It is the guard bands, which are unrecorded areas of the tape, that are the cause of the most objectionable noise bars.

The noise bars can be heavier when a tape recorded with a deck that uses narrow heads is played back even on a machine that uses

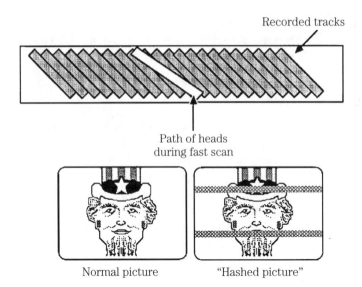

Recorded tracks

Path of heads
during fast scan

Normal picture        "Hashed picture"

■ **2-23** *The path of the video head during fast-scan playback crosses over several recorded tracks, resulting in a banded or* hashed *picture.*

wide SP heads. The bars are thicker because the wider heads pick up extra guard band area.

## Special three- and five-head systems

A number of older model VCRs have an extra third or fifth head that is used for noiseless freeze-frame or slow motion (three- and five-head VCRs are seldom made these days). The extra head, which is often referred to as the "trick" head, is not used for recording and normal playback, but only to deliver the cleanest possible picture in freeze-frame and slow-motion operation. Figure 2-24 shows the geometry of three- and five-head systems and how the heads are used for the various operations.

This extra head can be a source of confusion during cleaning and maintenance. If the playback heads become dirty with tape oxide buildup or other foreign matter, you won't see a picture. But should you place the deck in freeze-frame or slow-motion mode, the picture might return. The reason for this is obvious: different heads are being used for standard and special-effects playback. Always be sure to thoroughly clean all of the heads before assuming that something serious is wrong with the VCR.

## Six-head VCRs

A growing trend among some manufacturers is incorporating six video heads, rather than two or four. The function of the additional

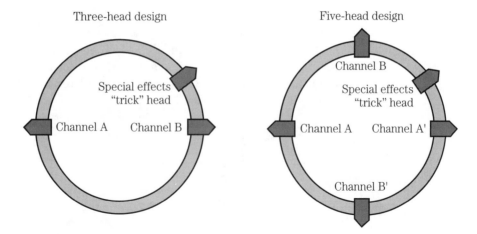

Three-head design

Five-head design

Special effects "trick" head

Channel A    Channel B

Channel B

Special effects "trick" head

Channel A    Channel A'

Channel B'

■ **2-24** *Special three- and five-head designs used in some VHS decks.*

two heads is solely for improving the picture playback quality of EP (also called SLP) recorded tapes.

On a four-head VCR, two of the heads are used for recording and playing SP tapes; the other two are for recording and playing EP tapes. In addition, the four heads are used in differing combinations for special-effects operations at both SP and EP speeds. Because the heads perform double-duty, there is some compromise in their design. Conversely, on a six-head VCR, the primary four heads are used in the same manner. The two additional heads are optimized solely for record and playback at EP speed. The result, say the manufacturers, is better EP picture quality, though for most six-head VCR models, the improvement is only minor.

## Flying erase heads

In conventional VCRs, the full-erase head is used to totally "scramble" the magnetic coating on the tape so a new signal can be properly recorded. Without full erasure, a portion of the old program might still remain. You might hear a faint background sound in the audio portion, and the video could appear washed out, grainy, or contain wavy, colored lines (called *moiré*).

VCRs with video dubbing capability use the video heads to rerecord over old programming. The results are acceptable but marginal. A flying erase head is mounted just before the video head in the head-drum assembly (see Fig. 2-25 on the next page) and erases just the video tracks and prepares them for proper rerecording. With flying erase heads, the video dub looks better and is not marred by moiré patterns. As shown on the next page.

Flying erase heads are often found in camcorders because they are used for start-and-stop shooting. The flying erase head provides for clean, glitch-free transitions between scenes, as shown in Fig. 2-26, even if you record both picture and sound over a previous segment. In most consumer VCRs, the flying erase head deletes two video tracks at the same time.

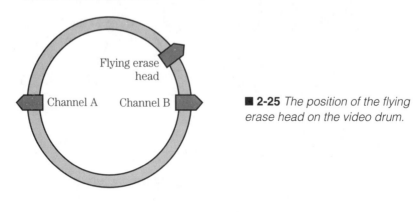

Flying erase head

Channel A    Channel B

■ **2-25** *The position of the flying erase head on the video drum.*

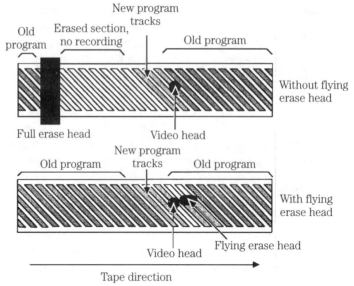

New program tracks

Old program    Erased section, no recording    Old program

Full erase head    Video head    Without flying erase head

New program tracks

Old program    Old program

With flying erase head

Video head    Flying erase head

Tape direction

■ **2-26** *The effects of insert editing recording with and without a flying erase head. The full erase head causes a large gap of unrecorded material between old and new programs; the flying erase head leaves no such gap.*

## Head switching

During recording, the signal feeds alternately to the heads via a head-switching circuit. The head is turned on as it touches the tape and makes its half-circle pass around the head drum. Just before the head is about to leave the tape, it turns off and the signal is applied to the opposite head. The video waveform for head switching is shown in Fig. 2-27.

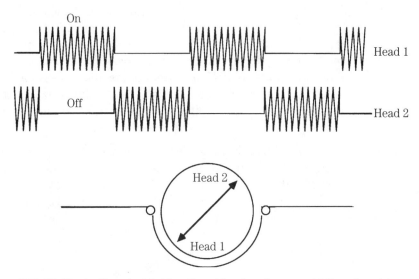

■ **2-27** *The action of switching the video heads on and off as the video drum rotates.*

## Inside hi-fi audio

The first Beta and VHS VCRs recorded monophonic sound only. In 1982, Sony came out with a novel stereo system for their Beta decks that recorded high-fidelity stereophonic sound. Not to be outdone, JVC soon developed a stereo hi-fi system for their competing VHS decks.

Although both Beta and VHS hi-fi decks share similar audio specifications, the process of recording and playing back the audio is completely different between the two systems. Both VHS and Beta hi-fi offer a dynamic range (soft to loud) of about 80 dB. Conversely, the linear audio track on conventional VCRs delivers a dynamic range of no more than 50 dB.

### Beta hi-fi

VCRs use rotating video heads to record large amounts of information in a relatively short space of tape. In a VCR, the tape literally crawls along, but the helical scan system enables the video heads to trace a far greater surface area of the tape. Similarly, high-fidelity sound is not possible using the stationary audio heads in the VCR because the tape moves too slowly. VCR engineers reasoned that if they improved video performance by using rotating video heads, they could do the same thing by using rotating audio heads.

Using a second set of heads would be the expensive way to add high fidelity to a VCR. Therefore, the hi-fi sound is included with the video and recorded with it on the tape. Figure 2-28 shows the standard Beta recording format, discussed in this chapter, and the Beta hi-fi recording format. Notice two changes: the luminance carrier is moved up 400 kHz (in Super Beta decks, discussed in a following section, the luminance channel is moved up 800 kHz). That opens more space between the luminance sideband (the diminishing harmonics of the luminance carrier) and the chrominance carrier.

The Beta hi-fi audio signal is frequency modulated and placed in the nook between the luminance sideband and chroma carrier. The sound is modulated onto FM carriers so that for one video field, the frequencies are 1.38 MHz (left channel) and 1.68 MHz

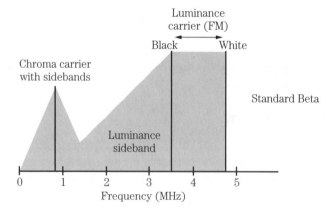

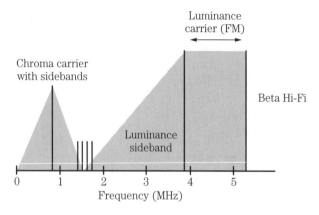

■ **2-28** *Signal layout for Beta standard and hi-fi recording formats.*

(right channel). During the next field, the FM carriers are placed at 1.53 and 1.83 MHz. The difference in frequencies between the fields helps reduce crosstalk.

Although the Beta hi-fi system moves the recorded signal around a bit, tapes recorded in hi-fi are still playable on most decks that lack the hi-fi capability. A monophonic rendition of the sound is placed on the audio track and the deck seeks out the luminance carrier and locks onto it, even though it is 400 Hz higher than it should be.

## VHS hi-fi

Beta was first to market a hi-fi VCR. JVC had considerable trouble developing a competing hi-fi audio standard for their VHS decks. The standard VHS recording format does not allow enough space between the luminance sideband and the chrominance carrier, even if the luminance carrier could be moved up a few notches. In fact, VHS decks cannot tolerate a change in frequencies.

This limitation means that VHS VCRs cannot use the video heads to record hi-fi audio, as Beta VCRs do (see the previous section on Beta hi-fi). A second set of heads, as shown in Fig. 2-29, are added to the video-head drum. The heads operate in the same manner as video heads. One head in the set records a field's worth of audio information, the next head records another, and so forth. The gap of the hi-fi heads can vary from about 26 microns to 42 microns.

In the VHS hi-fi system, the audio signals are recorded first, then the video signals are recorded. Normally, overrecording with the video heads would lead to complete or partial erasure of the audio

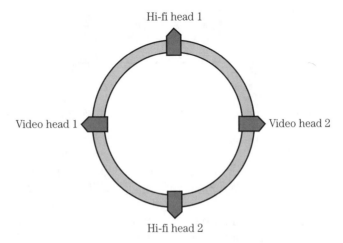

■ **2-29** *The position of the video and audio hi-fi heads on a typical VCR.*

track, but in practice, this does not happen—thanks to an old recording technique.

It's been known for many years that low-frequency signals penetrate into the tape farther than high-frequency ones. Because the hi-fi audio portion is lower in frequency than the 3.4 to 4.4 MHz luminance carrier, it travels deeper into the tape. JVC calls this technique *depth multiplex recording*. It sounds complex, but it isn't.

During recording, the hi-fi heads record their signal a split second before the video heads record theirs. The audio signal goes deeper into the tape, so when the video heads pass by, most of the already-recorded sound information remains. To avoid crosstalk between the audio and video carriers, the hi-fi audio heads are tilted at ±30°. The extreme azimuth also reduces crosstalk between right and left stereo channels.

The hi-fi heads record the audio on FM carriers, as shown graphically in Fig. 2-30. The frequencies are 1.3 MHz for the left audio channel and 1.7 MHz for the right audio channel. Compatibility with decks that do not feature hi-fi is retained because the sound also is recorded on the linear audio tracks. Some stereo hi-fi VCRs also have linear stereo audio heads.

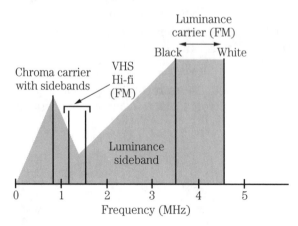

■ **2-30** *Signal layout for the VHS hi-fi recording format.*

## Hi-fi limitations

Because the hi-fi audio is recorded with the video track, effects such as audio dubbing are not possible. There is no way to selectively erase the audio portion of the program and replace it with another track. However, many VHS hi-fi machines also record stereo audio onto one or two linear audio tracks. These tracks can be selectively erased and rerecorded.

# Super Beta

Super Beta adds another 20 percent detail to the video picture. It does this mainly by relocating the luminance carrier (some additional circuitry also is used). Instead of positioning the luminance carrier between 3.5 and 4.8 MHz (a 1.3 MHz deviation), as in standard Beta units, Super Beta decks translate the luminance carrier up 800 kHz to 4.4 through 5.6 MHz (a 1.2 MHz deviation). Resolution is increased from 250 horizontal lines (standard Beta) to about 290 horizontal lines (Super Beta).

Compatibility with most older Beta decks that lack Super Beta circuitry is maintained because all Beta decks can seek out the higher luminance carrier frequency and lock onto it. However, some early-model Beta decks have been found that do not successfully track the higher frequency.

# Super VHS

VHS VCRs cannot cope with deviations in frequency standards, so to apply the same technique used in Super Beta with VHS units would impair compatibility. Decks that lack the Super VHS circuitry would not be able to play back tapes recorded on a Super VHS deck.

JVC finally took the plunge in 1987 and revealed a Super VHS deck. It uses the same general technique found in Super Beta decks, plus it adds a few other capabilities as well. From the outside, a Super VHS deck looks the same as a conventional VHS video recorder and has similar features such as three-speed recording (most models), wireless remote controls, fast scan in reverse and forward, and freeze-frame.

However, inside are electronics that record the picture in a wider bandwidth of frequencies. The luminance carrier for Super VHS is placed between 5.4 and 7 MHz, a deviation of 1.6 MHz. This is opposed to the conventional VHS decks, as shown in Fig. 2-31 on the next page, that have a 1 MHz luminance carrier deviation between 3.4 and 4.4 MHz. As shown on the next page. The wider carrier bandwidth, coupled with the shift to the higher frequency range of 5.4 to 7 MHz, provides the Super format with higher resolution—more than 400 horizontal lines compared to 250 lines in a conventional VHS deck.

Another constituent of Super VHS is separate luminance and chrominance signals. In a conventional VCR, these signals are blended together when recorded on tape and processed by the TV,

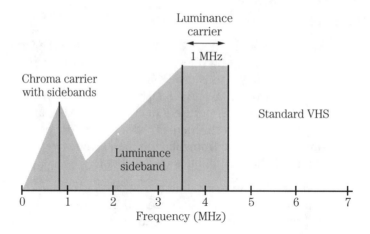

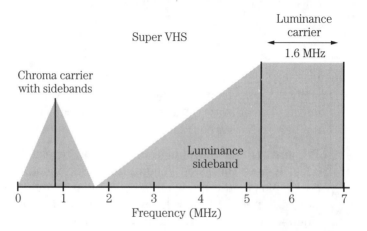

**■ 2-31** *Signal layout for VHS standard and Super recording formats.*

but they are separated before you see the picture. In a Super VHS deck, the luminance and chrominance channels are not mixed but are available at separate output terminals. The separated luminance and chrominance signals reduce crosstalk. The crosstalk appears as a waving herringbone or rippling effect you sometimes see when a tight, regular pattern is displayed in the picture.

Super VHS decks come with three types of video connectors for attaching the machine to your TV: RF, composite video, and Y-C. You get the lowest image sharpness when viewing the picture through the RF terminal, and slightly better quality—about 350 lines of resolution—when watching through the composite video terminal. Your set needs a composite video input on the TV to accept the signal. The full 400-plus lines of video resolution are available only when using the Y-C outputs of the Super VHS deck. Of course, you need a matching set of terminals on your TV.

Y-C connectors are also used in ED Beta, a professional Beta format used for very high quality recordings. Video equipment makers have begun to add Y-C connectors on other gear, including video routing switchers and laser disc players.

Note, however, that video manufacturers have greatly overplayed the importance of the Y-C connectors. In all video transmission techniques, the luminance and chrominance signals are mixed together. A comb filter separates the signals into their constituent parts. Comb filters are used in VCRs for recording the signals and in TVs and monitors for displaying the picture. In Super VHS and ED Beta decks, the Y-C outputs serve simply to separate the luminance and chrominance before they reach the TV. This eliminates redundant combining and reseparating of the Y-C signals.

Although the Y-C connectors on a Super VHS or ED Beta deck and television can make a difference in picture quality, they are not a universal panacea. In a technical paper distributed electronically via the Compuserve computer information network, engineer Bill Rood describes the pros and cons of using Y-C connectors when attaching various types of equipment, as shown in Table 2-2. Rood notes that although it won't hurt anything to connect, say, a TV tuner to a Super VHS deck via the Y-C connectors, it won't improve the picture quality.

■ Table 2-2 Using Y-C connectors.

| From | To | Use Y/C? |
|------|-----|----------|
| Super VHS | Super VHS | Yes |
| Super VHS | ED Beta | Yes |
| Super VHS | Y/C monitor | Yes |
| ED Beta | ED Beta | Yes |
| ED Beta | Super VHS | Yes |
| ED Beta | Y/C monitor | Yes |
| Laser disc player | Super VHS | No |
| Laser disc player | ED Beta | No |
| Laser disc player | Y/C monitor | No |
| TV tuner | Super VHS | No |
| TV tuner | ED Beta | No |
| TV tuner | Y/C monitor | No |
| Satellite receiver | Super VHS | No |
| Satellite receiver | ED Beta | No |
| Satellite receiver | Y/C monitor | No |

The maximum resolution of tapes you see on a Super VHS deck really depends on the quality of the original signal. If the original is low quality—taped off the air or through cable for example—the resolution will be the same as with a conventional VCR, even if you use a TV that is Super VHS ready. To get full resolution, you must record from a high-quality source like a videodisc, play back a prerecorded Super VHS movie, or make Super VHS tapes with a camcorder.

Because Super VHS uses higher frequencies to record images, tapes made on a Super deck can't be played back on a conventional VCR, even though both machines use VHS-format tape cassettes. However, a Super VHS deck can play back and record in conventional VHS format, allowing you to share tapes with friends and family and play commercially prerecorded tapes on your Super deck. But when you try to play a tape recorded with a Super VHS deck on a conventional VCR, all you see is snow.

Tapes made for conventional VCRs are not formulated for the high frequencies used in Super VHS recording; you need an extra high grade cobalt-doped ferric-oxide tape to get good pictures. As expected, the tape is more expensive than standard grade VHS tapes.

A small pinhole is drilled in the bottom of the Super tape cassette, enabling the deck to distinguish between grades. If the deck senses there is no hole, it shifts into standard VHS mode, recording and playing back tape like a conventional VCR.

The Y-C cable consists of two pairs of shielded wires for a total of four wires. You can make your own Y-C cables using good-quality shielded wire and the proper size DIN connector. The pin assignments for the connectors are:

Pin 1    Y ground
Pin 2    C ground
Pin 3    Y signal input/output
Pin 4    C signal input/output

Note that a cable designed for use with the Apple Macintosh computer uses the proper size DIN connector for the Y-C cable, but the internal connection of the wires is not compatible. You cannot reliably use the Apple cable on your video system.

## Digital effects

Push the PAUSE button on a VCR equipped with digital-effects circuitry and the picture is sharp and clear—not noisy and jittery as with most home VCRs. Why such crystal clarity? With conventional VCRs, still frames are made by stopping the tape and having the rotating video heads scan the same TV image tracks. The result is jitter, rolling, and noise bars—bands of salt-and-pepper flecks—particularly at the top or bottom of the screen. Digital VCRs use computer *RAM* (random-access memory) chips to store a digitized image of the picture. Thus, you see the still frame frozen in silicon.

Having clean still pictures is only one benefit of the newest digital VCRs. Depending on the model, you also get noiseless slow-motion and fast-motion playback, digital enhancement of the picture during normal playback, unique special effects such as posterization and mosaic, picture-in-picture effects, and more.

The term *digital VCR* is really a misnomer. *Digital-effects VCR* would be a more accurate term because the digital circuits use video special effects during playback, but the VCR does not actually record digital data on tape and then play it back. The recording and playback medium is still analog, as with conventional VCRs. The digital circuits are primarily used for special effects, such as freeze-frame, slow motion, and picture-in-picture (also known as *PIP*).

Digital video effects are achieved by routing the analog playback signal (coming from the tape or through the VCR tuner) to an *A/D* (analog-to-digital) converter IC. This chip transforms analog signals into their digital counterparts. The digital information is then processed and stored in dynamic RAM in much the same way as a memo or spreadsheet is stored in the memory of a personal computer. Digital VCRs have from 64 to about 300K (kilobytes) of RAM packed inside them, or about the same as the average desktop computer. Before you see the final picture, the digital data stored in RAM is reconverted using a *D/A* (digital-to-analog) converter.

Most current digital VCRs use 6-bit digitization, so any picture element, or pixel, on the screen is stored as a 6-bit binary number in RAM. Six bits results in a maximum of 64 possible combinations, which is what causes a slight digital look or *blockiness* during playback. With more bits representing each picture element, there would be more resolution in the final picture and it would look more uniform.

Once in digital form, the picture data can be manipulated to create unusual special effects. For the posterization and mosaic effects built into some of the digital-effects models, the data is read out in different ways.

Digitally enhanced slow motion is another common feature to digital-effects VCRs. With most decks, the slow-motion rate is preset, but on others you can vary the slow-motion speed. Many digital-effects VCRs have a frame-advance feature that lets you step through still frames one at a time. Frame advance is like slow motion that you control manually. The frame advance can be coupled with picture-in-picture effects to show a sequence of frames on the screen.

Some digital-effects VCRs have a PIP feature. With PIP, a smaller picture from the VCR tuner or the tape is displayed in one corner of the screen. Controls on the deck let you select which image you want displayed full size (you only get audio for the full-size picture) and which corner you want the PIP image to appear.

Digital circuitry also can be used for reducing noise in the playback picture, but as of this writing, only a few models have it. It works as follows. Each video frame is composed of two fields. The lines from the two fields interlace with one another. During playback, the VCR digitizes the first field. The information from the first field is then averaged with the information from the second field. The averaging acts to cancel out random noise, so the picture looks sharper.

## Time-base corrector

A *TBC* (time-base corrector) is a device that corrects the hiccups in timing and synchronization that naturally develop in a television signal. For decades, the Federal Communications Commission has required television stations to feed their broadcast signal through a time-base corrector before that signal is beamed into people's homes. The result is a more stable picture.

Recently, a small handful of consumer video products have been endowed with time-base corrector circuitry. Although home videos seldom require time-base correction, the improved image stability makes for better-quality tapes. Time-base correction in home videos is particularly beneficial when copying from VCR to VCR, like when making copies of Junior's first birthday tape for Grandma and Auntie.

The TBCs used in television stations and those stuffed into consumer VCRs and videodisc players use similar principles of design, but they are not the same. The time-base correctors currently found in consumer gear use only one or two of the features normally found in an industrial TBC device. This selectivity is understandable: a time-base corrector designed for professional use costs $3000 to $4000—many models are priced much higher. Conversely, the few VCRs that use TBC circuitry retail in the neighborhood of $1500. That price also pays for other top-model features such as Super VHS, digital special effects, and jog/shuttle tape transport.

Exactly what does a TBC do, and how does it function? Studio-grade time-base correctors perform three major jobs:

- ☐ Find the sync pulses in a signal and rebuild them according to rigid specifications.
- ☐ Retime each of the 525 lines that comprise a complete video picture so the lines start and stop at exactly the right spots.
- ☐ Monitor and alter the relative strength of certain signals including those that control color and synchronization.

Recall from earlier in this chapter the format of a television signal. The beginning of the video signal is comprised of synchronization pulses that control the vertical hold of your TV. Immediately following the vertical sync pulses are the individual lines that make up the TV picture. Each of these lines has a specific duration, and each starts with a short pulse to control the horizontal hold on your TV.

The vertical and horizontal synchronization pulses form the backbone of the TV signal. Without these pulses, no recognizable image would appear on the TV tube. During recording, playback, and transmission of a television signal, these synchronization pulses can become warped and distorted, and that can affect the stability of the picture. For example, if the vertical pulses are missing or severely distorted, the picture could twitch or jitter.

The first job of a time-base corrector is to detect and strip out the old synchronization pulses. The old pulses are then replaced with fresh ones. The rejuvenated sync helps to stabilize the picture so it can be viewed without annoying rock and roll.

The second job of a time-base corrector is to make sure all the picture lines start and stop at the same time. This is where the time-base correction comes in. If any of the lines start too soon or too late, the picture appears squiggly and uneven. The symptom is called *time-base error*. The effect is most noticeable when the screen shows a vertical object, like a telephone pole or fence post. With time-base error, the individual horizontal lines of the TV picture are shifted left and right in a random pattern.

A time-base corrector samples each line of video and times out its duration. The TBC knows that each line of video is supposed to be exactly 63.5 microseconds (63.5 millionths of a second) in length. If a line is shorter, the next line will start sooner, and the image will shift toward the left side of the screen. If a line is longer, the next line will start later. In turn, the image will shift toward the right side of the screen.

At the heart of the time-base corrector is a video line "holding tank." This holding tank memorizes one or more lines of video, then replays them at precise intervals. On coming out of the time-base corrector, each line is exactly 63.5 microseconds in length—no more, no less. The problem of lines shifting left and right on the screen are now gone.

Professional-quality time-base correctors memorize a complete frame of video—525 lines—at one time. Each line of video is digitized and stored in memory until its turn is up. On leaving the memory bank, the signal is reconverted to analog form. The conversion between digital and analog together with the copious amounts of memory required to store the signal require hefty and expensive circuitry, contributing to a large portion of the high price of a studio-grade TBC.

The third job of the time-base corrector is to selectively modify the intensity and characteristics of certain components of the television signal. One of these components is the *color-burst* signal, which provides color to an otherwise black-and-white image. Turn a dial on the time-base corrector and it modifies the color burst so greens become blue, blues become red, and so forth. With a twist of another dial, the intensity of the colors are varied—from a pale pink to a vibrant red, for instance. These and other changes are sometimes necessary if the original signal is weak or has been severely degenerated.

The ideal video signal doesn't need processing with a time-base corrector because there's nothing to correct. But as video pulses are recorded on tape and pass through mazes of cables, the signal becomes degraded. Picture quality is chiefly reduced as the video signal is recorded from one VCR to another. With each generation of recording, the original signal becomes more and more garbled until little is left of it and the picture becomes a mass of unwatchable jitter. A time-base corrector used during rerecording helps restore the signal to its original quality. It's important to note that a time-base corrector doesn't touch the actual video portion of the signal, just the synchronization pulses and timing of that signal.

Of the three basic jobs of a time-base corrector, the current crop of TBC consumer video gear does just one task: corrects for horizontal-line time-base error. On most decks, only one or two lines of video are memorized at a time. This limits the "window of correctability" to just a few microseconds in either direction. For example, if a line is 63.0 microseconds—instead of the normal 63.5 microseconds—the TBC circuits will stretch the line 0.5 millionths of a second, restoring it to its proper length. But if the line is 60 or 61 microseconds in duration, the timing error is excessive and the TBC circuits won't be able to compensate. Fortunately, time-base error of this magnitude is rare—unless the tape is damaged or the VCR is severely out of whack.

## Elapsed-index or real-time counter

Conventional VCRs, especially those made up until 1989 or so, used elapsed-index counters to denote the relative start and stop points of a program recorded on a tape. Zero the counter and it reads "0000." Start the deck, and the counter moves up, ticking off one digit for every couple seconds of recording time.

The problem with elapsed-index timers is that they are inherently mechanical, even if the counter display is electronic. As you will learn in later chapters, the counter is driven by a sensor located in the take-up reel inside the VCR. As the reel moves, the counter advances.

As with any mechanical counting technique, errors are bound to crop up. Because the design of VCRs vary, elapsed-index counters show only relative tape movement. Just because your favorite episode of "My Favorite Martian" starts at counter position 0654 on one VCR doesn't mean it will start at the same spot when played back on another VCR.

A real-time counter, on the other hand, uses the control-track pulses recorded with the VCR to accurately display the elapsed time since the tape was first started. The counter ticks off in seconds, minutes, and hours. As long as you remember to reset the counter at the beginning of the tape and the tape is not otherwise damaged, you can return to any spot just by noting the elapsed time. When using a VCR equipped with a real-time counter, you will notice that the counter doesn't advance when playing a blank portion of tape. If the tape is fresh and has never been recorded on, it will lack a control track; hence, the real-time counter won't be advanced.

## Autotracking

One of the banes of using a VCR is fine-tuning the picture using the tracking control. This control compensates for the inevitable difference between VCRs, even when using the same VCR to play back an older tape. Without the control, snow could often appear on the top or bottom of the screen.

The 8mm decks and camcorders incorporate an ingenious automatic tracking system, so there's no need to adjust the tracking control. Tracking controls are common on Beta and VHS decks, although a number of the later-model VHS units incorporate an auto tracking feature. In the VHS automatic tracking system, electronics in the deck sense when the video signal drops 15 percent or more during playback. This drop indicates the tracking is off, thereby causing snow at the top or bottom of the picture.

## Hi-band 8mm

Akin to Super Beta and Super VHS is Hi-band 8mm (also called *Hi8*). The 8mm format is primarily limited to camcorders, and this is where Hi-band finds its most popular use. As with the other Super format, Hi-band 8mm enjoys an improved picture due to an increase in the luminance bandwidth. Proponents of the Hi-band system claim an increase of picture quality from 30 to 50 percent; the improvement is particularly noticeable when copying the tape.

## Fuzzy logic

Video makers have recently introduced a unique form of digital circuit called *fuzzy logic* to camcorders and home decks. Although "fuzzy logic" is a funny name, it's the result that matters. The chief benefit of fuzzy logic is that, unlike other computer circuits, approximations are acceptable. In a computer circuit, there are only two states: off and on (or 0 and 1). In fuzzy logic, there's a "maybe," or more precisely, graduations between the logical 0 and 1 states.

Because approximations are tolerated, fuzzy logic circuits perform a type of artificial intelligence akin to the way you make educated guesses. So far, the main applications for fuzzy logic is in such areas as camcorder focus control, brightness adjustment, and image stabilizing. However, the technology can expand to all areas of video.

# PCM audio

All 8mm decks can record high-fidelity digital sound. The specifications are almost as impressive as those found on audio compact discs: 50 Hz to 15 kHz frequency response, 90 dB signal-to-noise ratio, and 88 dB dynamic range. The analog-to-digital conversion uses 8-bit quantization (by comparison, compact audio discs use superior 16-bit quantization).

The process of digital recording is known as *PCM* (pulse-code modulation). In operation, a series of pulses that represents digital data is recorded on the tape. The pulses are derived by converting the original analog audio signal to digital form. For playback, the pulses are processed in just the reverse to reconstitute analog audio from the digital.

There are two digital audio modes: video-plus-audio and all-audio (the latter is sometimes called *multi-PCM*). In the video-plus-audio mode, the digital sound is recorded with the video picture (the track also can be added after the picture is recorded). In all-audio recording, the deck records only sound. Depending on the recording speed, you can store up to 24 hours of digital sound on a 2-hour tape (12 hours at SP speed). Figure 2-32 shows the track layout for video-plus-audio and all-audio recording.

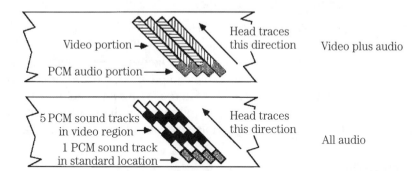

■ **2-32** *Tracks recorded on 8mm tape in the video-plus-audio and all-audio modes.*

Note that in video-plus-audio mode, the PCM audio is placed at the head of each video track. This is accomplished by wrapping the tape farther around the head drum, as depicted in Fig. 2-33 on the next page. The standard 180° wrap is used for video recording, and the extra 30 degree is used for PCM audio recording. PCM audio also is found on some high-end VHS decks. Most use higher-quality 14-bit quantization.

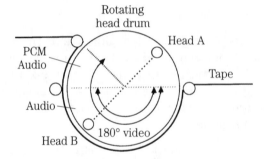

**■ 2-33** *The tape is wrapped around the head in the 8mm system an extra 30 degrees to accommodate the PCM audio track.*

## Cable box control

Many of the latest mid- and high-end VHS VCRs come with a feature to remotely control your cable box, if you have one (the cable box must be equipped with an infrared remote control sensor, and the remote control capability activated). This features entails the use of an external cable box controller. The controller attaches to the VCR—usually via a plug in the back of the deck—and fits over or near the cable box.

The controller is designed to make it possible for the VCR to control the channel selection on the cable box, so you use the programming features of the VCR to record shows while you're away. The VCR starts and stops recording, and uses the cable control box to change the channels on the cable remote control.

The cable box control is an interim effort to make it easier for a VCR to remotely command a cable box. The cable box control is provided with VCRs in answer to a new law that requires better interconnections between cable equipment and home video equipment. Eventually, manufacturers agree, VCRs and cable boxes will have connectors on them that will accept a direct connection between the two. The use of a cable box control device will not be required.

## Controls

Most VCRs have the following typical controls, as shown in Fig. 2-34 on page 73:

- ☐ POWER   Turns the power on and off.
- ☐ EJECT   Ejects the cassette from the VCR.
- ☐ PLAY   Places the deck in the PLAY mode.
- ☐ FAST FORWARD   Fast forwards the tape; no picture on screen.
- ☐ REWIND   Rewinds the tape; no picture on screen.
- ☐ CUE (or forward fast scan)   Scans the tape at 3 to 30 times the normal speed in forward; high-speed picture on screen.

- [ ] REVIEW (or reverse fast-scan)   Scans the tape at 3 to 30 times the normal speed in reverse; high-speed picture on screen.
- [ ] PAUSE   Stops the tape during playback; a still frame appears on the screen (the deck stops automatically after about five minutes to prevent excessive tape and video-head wear).
- [ ] RECORD   When used with the PLAY button, places deck into RECORD mode.
- [ ] AUDIO DUB   Records just audio over previously recorded program (video portion must be previously recorded).
- [ ] VIDEO DUB   Records just video over previously recorded program.
- [ ] TRACKING   Adjusts tracking to reduce or eliminate noise bars in picture (not found on 8mm decks).
- [ ] OTR   One-touch recording; starts deck in RECORD mode and records for preset period. On many decks, each press of the OTR (or EXPRESS) button increases the recording time by 30 minutes.
- [ ] TIME buttons   Set the record timer.
- [ ] CLOCK buttons   Set the clock (many decks set the time and day of week).
- [ ] CHANNEL ¾ output (back of deck)   Switches RF modulator output to channel 3 or 4.
- [ ] LINE (with A/V terminals)   Switches between RF or direct AUDIO/VIDEO input terminals (might be automatic on some models).

Most VCRs have a counter and counter RESET. The counter is operated by the take-up spindle (either mechanically or electronically) or the control-track pulses and indicates relative tape position. The counter is reset to "0000" by pressing the RESET button.

## Connectors

VCRs contain a plethora of connectors located on the front, side, and back of the deck. Refer to Fig. 2-35 on page 73 for the typical connectors you're likely to see on the back of a VCR and what they do.

- [ ] VHF IN   The RF connector for receiving channels 2 through 13. Two sets of connectors can be used—one for 300 Ω twinlead and the other for 75 Ω coax.
- [ ] UHF IN   The RF connector for receiving channels 14 through 83. One set of connectors is used for hooking up to 300 Ω twinlead.

☐ VHF OUT    The RF connector for hooking up to the VHF input of the TV. Two sets of connectors can be used—one for 300 Ω twinlead and the other for 75 Ω coax.

☐ UHF OUT    The RF connector for hooking up to the UHF input of the TV. One set of connectors is used for hooking up to 300 Ω twinlead.

☐ VIDEO IN    The video input to the VCR. An RCA phono connector for receiving baseband composite video from another source such as a video disc player, satellite receiver, or another VCR.

☐ Y-C OUT    The video output from the VCR. An RCA phono connector for hooking up to a TV/monitor, another VCR, etc.

☐ AUDIO IN    The audio input to the VCR. An RCA phono connector for receiving baseband audio from another source such as a video disc player, satellite receiver, or another VCR. Might have two connectors if the VCR is stereo.

☐ AUDIO OUT    The audio output from the VCR. An RCA phono connector for hooking up to a TV/monitor, another VCR, etc. Might have two connectors if the VCR is stereo.

☐ Y-C INPUT    The separate luminance and chrominance video inputs to a Super VHS or ED Beta VCR. A four-prong connector for receiving baseband luminance and chrominance signal from another Super VHS VCR or similarly equipped video source.

☐ VIDEO OUT    The separate luminance and chrominance video outputs of a Super VHS or ED Beta VCR. A four-prong connector for hooking up to similarly equipped TV/monitor, another Super VHS VCR, etc.

☐ MICROPHONE    The microphone audio input jack, which usually accepts a ⅛-inch miniature plug. Might have two jacks if the VCR is stereo.

☐ HEADPHONE    The headphone jack, sized for either ¼-inch standard or ⅛-inch miniature plugs. Stereo VCRs will have a three-way jack for use with stereo headphones.

☐ MTS DECODER    The multipin connector for hookup to an outboard MTS stereo decoder. The type of connector will vary depending on make and model.

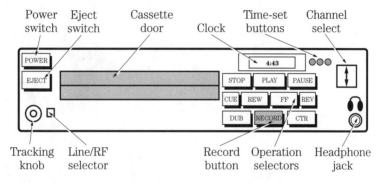

Power switch Eject switch Cassette door Clock Time-set buttons Channel select

POWER EJECT

4:43

STOP PLAY PAUSE
CUE REW FF REV
DUB RECORD CTR

Tracking knob Line/RF selector Record button Operation selectors Headphone jack

■ **2-34** *The front-panel controls on a typical VCR.*

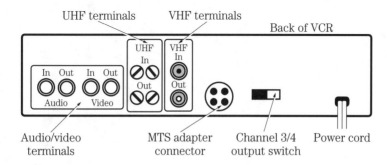

UHF terminals VHF terminals Back of VCR

In Out In Out
Audio Video

UHF In Out
VHF In Out

Audio/video terminals MTS adapter connector Channel 3/4 output switch Power cord

■ **2-35** *The back-panel connectors on a typical VCR.*

73

# The VCR environment

SUCCESS WITH YOUR VIDEOCASSETTE RECORDER DEPENDS ON how well it is integrated with the rest of your video system. A haphazard arrangement will yield inferior results, and you won't be able to enjoy the full capabilities of the deck. Many potential problems can be completely avoided by proper placement and hookup of your VCR.

If your deck is giving you problems, check this chapter first. It provides details on proper installation, adjustment, and use of a VCR. Home video isn't a difficult technology to use; on the contrary, it's simple. Nevertheless, a VCR is unlike the trusty TV set, and if you've never owned one before, you might be surprised by the special considerations that you need to keep in mind.

This chapter also details some extras you can add to your video system to improve the picture and sound and therefore the enjoyment of your video investment.

## Unpacking

If you have yet to buy or unpack your VCR, here are some quick tips to keep in mind.

### Concealed damage

When unpacking your VCR (especially one you bought by mail order), look for concealed damage. Concealed damage is breakage that you can't see on the outside of the box but becomes apparent when you take the deck out of its shipping container. If the machine has been damaged, return it immediately to the dealer. Exterior damage is a good indication that internal components are damaged.

Save the box and all packing material, at least until you've had the deck a month or so or until the warranty is up. Most mechanical and electrical problems will arise in this time, and you'll need the box to return the deck for warranty repair.

### Check packing list

Most VCRs come with a variety of accessories, including cables, cable adapters, and a remote control. Verify the packing list against the contents of the box to make sure you have everything you are entitled to receive. If you have purchased a camcorder, the unit might come with a cassette adapter (if it is the compact VHS variety), a battery, battery recharger, and carrying case. Check the manual to be sure that the accessories are present. If any are missing, consult your dealer.

## Installation

Keep the installation tips in this section in mind when adding a VCR to your video system.

### Ventilation and video furniture

VCRs generate a certain amount of internal heat, so it is important to avoid obstructing the ventilation slots on the top, sides, back, and bottom of the deck. A machine that gets too hot will perform erratically and can be damaged. Avoid placing the deck on the top of the TV set, a place that normally gets fairly hot after an evening of watching the nighttime soaps.

One way to provide adequate ventilation is with a piece of well-designed video furniture. Video furniture can not only improve the look of your living room or TV den, but it also helps protect your equipment. The least expensive is the roll-around TV and game-machine cart. Bookshelf furniture has more space for VCRs and accessories, but the cost is usually higher. Most bookshelf units come with wheels or rollers and have adjustable shelves. Large credenza furniture can hold a 19-inch or larger TV, VCR, disc player, satellite receiver, tapes, and more. Many have glass fronts and pull-out drawers.

If the furniture you like doesn't have any drawers for small accessories or tapes, consider purchasing a caddy or box that can be filled with these miscellaneous goodies. Be sure the tapes can stand on end (as described later). This position lessens the possibility of tape stretching.

When deciding on video furniture, keep in mind that you need to allow at least an inch or two of extra space on all sides—front, top, sides, and rear. This space not only allows for easier cable routing and better ventilation, but it lets you install and adjust your equipment with greater ease. Measure each piece of video gear you have, and add at least an inch to every dimension. Remember to include doors that open (front, side, or back access), cables that jut out the back, side, or front. Don't rely on the sizes given in the manual; the specifications don't take these important things into consideration.

Next, measure the inside of the furniture, and compare these dimensions to the sizes of your equipment. Be sure to measure the distance of the openings in the furniture. These can sometimes be smaller because of trim. Avoid making the installation permanent. Allow yourself the convenience of being able to modify the system.

You should always place the VCR on a level shelf or platform that's free of vibration and shock. Be sure that the shelf can adequately support the deck. If it looks like the shelf might break under the weight of the unit, by all means, find another place to put the machine.

## Remote control placement

You should place a VCR that has an infrared remote control in the open, yet out of direct sunlight. If possible, avoid placing it in an audio/video rack equipped with a glass front. The glass might disperse the infrared light beam and reduce the effectiveness of the remote. This precaution is especially important if the glass is smoked. You might need to keep the glass front open when operating the VCR with the remote.

Also, be sure that the deck is within range of the remote (usually 20 feet or less), and that you can aim the remote control transmitter in a direct line—more or less—to the VCR. The remote might not work if it is used at angles greater than 30 degree off to either side of the front panel of the deck, as shown in Fig. 3-1 on the next page.

# Hookup

Whether or not you plan to invest in video furniture, pay careful attention to how you hook up the pieces of your video system. By using the right type of cables, connectors, and accessories and wiring everything properly the first time around, you're assured of years of carefree operation. A messy installation only invites aggravation and poor-quality video.

## Proper cables

Have you ever taken a ride on a bumpy, country road? After the trip, you probably felt rankled and worn out. And your car! It'll

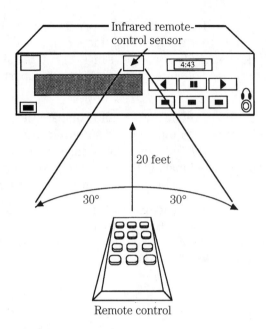

Infrared remote-control sensor

20 feet

30°  30°

Remote control

■ **3-1** *For best results, keep the wireless infrared remote controller within 30 degrees of the axis of the VCR's sensor and no more than 20 feet from the sensor.*

never be the same. Cables are the pathways for the signals generated by your video gear. If those signals have a rough time getting to their destination—that is, the television—the picture and sound quality definitely won't be pleasing.

That's why it's important you use the proper type of cable in your video system. The best cable for video is coaxial, which is made of a solid center conductor surrounded by a braided outer conductor.

## Types of coaxial cables

Three major types of coaxial cable (often referred to as *coax*) are used for video: RG-59, RG-11, and RG-6. The RG-6 and RG-11 varieties are different only in that the foam surrounding the center conductor is thicker than in RG-59. RG6 cable is the all-around best because it offers the greatest signal-passing capabilities for all channels and because it lasts longer under adverse conditions.

## Matched impedances

All three types of TV coax have impedances of 75 $\Omega$. Impedance, which is expressed in ohms ($\Omega$), is the measure of opposition an alternating electrical current encounters when sent down a cable or through a device. It's important that the impedance is matched throughout the entire video system. Ghosts and loss of signal can otherwise occur.

The impedance for the VHF input and output jacks on VCRs and TVs is 75 Ω. So the cables carrying those signals also must be rated at 75 Ω. There are other types of coax cables with different impedances, but they aren't meant for use with video. Stay away from cables designed for CB and amateur radio use (such as RG-8 and RG-58), because they have an impedance of 52 Ω. If you use them, they cause a signal imbalance that deteriorates the video signal.

## Twinlead

Another common type of TV hookup cable is *twinlead*, which is also used to carry VHF/UHF signals in a video system but isn't nearly as good as coax. Why? Coax provides superior rejection of noise and interference because the outside conductor forms a shield. Radio-frequency signals radiated from different parts of your video system (and nearby television stations) don't easily penetrate into coax because of the shielding. A video system using twinlead, especially twinlead that's weathered and cracked with age, is typically riddled with ghosts and interference.

Use coax cable with your video system as much as possible. If you have twinlead, outside or inside, consider changing it to coax. If you must mate coax with twinlead (the latter of which has an impedance of 300 Ω), use a matching transformer, discussed in a following section.

## The right connections

When hooking together coaxial cable, it isn't enough to twist the wires to form a connection. Coax cables need special fittings called *F* connectors (shown in Fig. 3-2 on the next page) on each end to ensure that the signal passes from cable to equipment without any loss.

The male end of the connector, built into the VCR, TV, or other component, is barrel-shaped with threads around it. The female end of the fitting is attached to the coax and can be either the thread-on or slip-on type. Take care when attaching an F connector onto the end of a cable. For information on how to do it, see appendix D, "Attaching an F connector." As an alternative, you can buy a ready-made cable with the F connectors already in place.

## Other types of cables

Keep in mind that coaxial and twinlead cables are meant for RF signals. For best results, you should use shielded video cable when hooking up components that pass baseband video and audio sig-

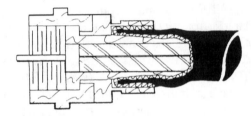

■ **3-2** *A cutaway of a coaxial cable and an F connector.*

nals—for example, the VIDEO OUT connector of a VCR to the VIDEO IN connector of a TV/monitor. Again, avoid using shielding cable designed for audio applications. This cable has the wrong impedance, and it isn't sufficiently shielded to reject interference.

You can buy video cable with RCA phono-type connectors on each end like the one shown in Fig. 3-3. This cable is the type used in consumer video gear that supplies a baseband input or output. Some high-grade baseband cables have a matched impedance of 75 Ω as well as gold-plated connectors. These contribute an extra measure of quality but aren't absolutely necessary.

■ **3-3** *RCA phono connectors.*

### Parts for proper hookup

One of the keys to assembling a top-notch video system is using the proper interconnecting components. There are several major component types, including matching transformers, band splitters, and A/B switches.

## Matching transformer

Matching transformers, also called baluns, match the impedance of one type of cable to that of another. They also provide the proper connector or terminal to mate unlike cables together. Figure 3-4 shows a transformer that adapts 75 Ω coax for connection to the VHF terminal on a television set.

■ **3-4** *A 300 ohm to 75 ohm balun (or matching transformer).*

Other matching transformers are available for going from 300 Ω twinlead cable or terminals (such as on an outdoor antenna) to 75 Ω coax. Matching transformers are available for both indoor and outdoor use. Outdoor transformers are waterproof to protect them from weathering.

## Signal splitter

Whenever you need to make one VHF/UHF signal go two or more different ways, you need a signal splitter. Splitters are small box-like attachments that have several fittings on them (see Fig. 3-5 on the next page). Splitters come in three popular styles: two-way, three-way, and four-way. These splitters are for splitting a signal two, three, and four ways, respectively.

Note that a signal splitter divides the strength of the signal as well, so you should limit the number of splitters in your system. A four-way splitter divides the strength half again for each output terminal, so you should not use a four-way splitter if you only need a two-way. If you must split the signal so many times that the picture looks grainy or washed out, use an amplified coupler (a splitter with an amplifier built-in) or an in-line coaxial amplifier.

■ **3-5** *A two-way signal splitter.*

In addition, avoid leaving one of the outputs of a splitter unused. If you can't fill the vacant spot, attach a terminating resistor to the connector, which can help eliminate ghosts.

### Band splitter

Band splitters take a combined VHF (channels 2 through 13) and UHF (channels 14 through 83) signal and split it into its separate VHF and UHF components (some even include a separate tap-off for FM radio). Band splitters are used with VHF/UHF antennas and some apartment cable systems. Some band splitters are designed as band combiners to enable you to mix VHF and UHF in one cable.

### A/B switch

Another common device you might need is the A/B switch, which is used in either of two ways: To route a signal to one of two different destinations, or to select between two different input sources. For example, you can set up an A/B switch (like the one in Fig. 3-6) with your TV so you can select between either of two VCRs. More elaborate selectors—ones that switch several signals two or more ways—are discussed later in this chapter. There are switchers for both RF and baseband audio/video signals.

### Attenuator

You might be lucky enough that the signals from the local TV stations or cable system are too strong. When the signals are too powerful, you often receive more than one channel at a time. The picture also can look very dark, have an excessive amount of contrast, or will roll and jitter. These effects are caused by overmodulation distortion.

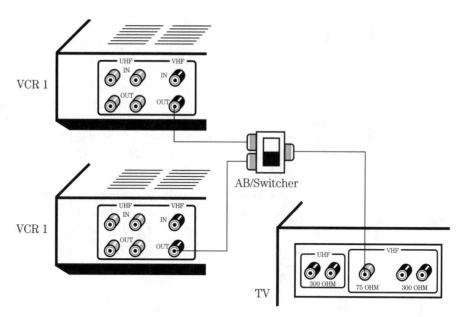

**■ 3-6** *One way to connect an A/B switch to route the signal between two VCRs to one television.*

*Attenuators* are used to weaken the signals reaching your TV set and VCR. Attenuators can be fixed or variable. A fixed attenuator reduces the signal by some known amount, usually 6 dB. A variable attenuator allows you to adjust the amount of signal reduction, usually from 10 to 20 dB. Variable attenuation is helpful if you receive strong nearby stations and weak, distant stations.

### Filters

There are several types of filters available for use with TVs and VCRs that reduce or eliminate interference from other signal sources. One common type is the FM filter, or trap, which removes interference caused by nearby FM radio stations. The problem can cause herringbone patterns on your TV when watching channels 5, 6, and 7. In installations where one cable carries both VHF and UHF channels to the TV, FM filters should be used in the VHF line after the VHF and UHF bands have been separated.

A highpass filter reduces the crackles and wavy lines caused by interference from local CB and ham radio sets. Static interference—pops in the sound and flashes on the screen—can be caused by such things as car ignitions and fluorescent lights. A TV interference filter can help cut down the static.

All of these filters attach between the cable or antenna and the TV. The ac interference filters attach to the power cord of your TV and help to block out annoying static caused by vacuum cleaners, blenders, hair dryers, and other motor-operated appliances.

### Signal amplifier

Signal amplifiers are used when you need to boost the signal coming from the antenna. Amplifiers are especially useful when the antenna is more than 100 feet from the VCR and TV. One type of amplifier attaches to the antenna mast and is powered by a small ac adapter that you connect into the coaxial cable leading to the VCR. The amplifier receives dc power through the center conductor of the cable.

When using such an amplifier, make sure there are no baluns between the amplifier and the power source. Most baluns block dc electricity, which would render the amplifier useless. Remember that it is always better (and usually cheaper) to replace the antenna with a better model than trying to improve reception by adding an amplifier.

Another type of amplifier, the amplified coupler, is intended for indoor use and is typically used to boost the signal so it can be adequately fed to various TVs and VCRs throughout the house. Amplified couplers plug directly into the ac outlet; they don't require a separate power adapter. A sample household wiring diagram, using an amplified coupler, is shown in Fig. 3-7.

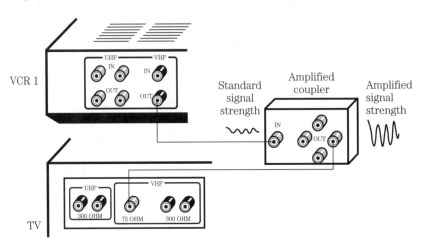

**■ 3-7** *An amplified coupler boosts the signal strength so that it can be sent through long lengths of coaxial cable.*

In-line coaxial amplifiers can be placed anywhere in the system and are powered in a similar fashion to mast-mount amplifiers. In-line amplifiers are ideal for boosting the signal when it must pass long distances to a remote TV or VCR.

84

### Grounding rod

Antennas and masts are made of metal, and because they stick up higher than most other objects around your house, they are susceptible to lightning. For your safety and the safety of your video gear, you should always ground the antenna with a grounding rod attached to a static discharge unit on the antenna. With this assembly, much of the dangerous static will pass from the antenna to the earth rather than into your living room.

Installing a grounding rod is simple: drive the rod firmly into the earth, then connect it using aluminum or copper grounding wire to a static discharge unit mounted on the antenna. For further protection, you also can install a grounding block anywhere along the antenna cable line feeding your VCR. For best results, locate the block outside where you can easily attach the ground wire to the grounding rod.

### Professional installation tips

Keep these points in mind when installing your video system:

**Plan on paper** Now that you've got the basics down, what next? When hooking up your system, it's best to place all your video gear—with nothing attached to them—in their proper places. It's a good idea to map out on paper how you'll connect your system before you do any of the actual work. Include all the cables, matching transformers, and splitters you'll need in your drawing. Take the time now to spot any problems. If you look hard enough, you can often find better and less expensive ways to connect your system.

**Actual hookup** When you're satisfied with your hook-up plan, turn off the power to your VCR and TV. Begin connecting the cables and components together starting where the signal is first introduced into your system; beginning just anywhere will lead to confusion if your system is of any size.

Use the shortest cable length whenever possible. If you have extra, don't wad it up into a ball or a tight loop. The solid center conductor in coax can be broken if you twist it excessively. Loop it loosely and bind the cable with a plastic tie wrap or twist tie. Label each end of the cable for future reference. Connect each cable the same way, making sure you arrange them to avoid tangling.

When hooking everything together, don't staple cables or wires to the inside or down the back of a cabinet or wall. You might puncture the insulation and cause some strange interference, or worse yet, cause a fire or shock hazard. Whenever possible, keep the ac cords separate from cables carrying video signals. Otherwise, you run the risk of video cables picking up interference and noise from the ac wires. If signal cables must cross ac cables, run them at 90° to one another, not parallel. Never place coax cables under carpeting, as walking over them can deform the insulation, which in turn, can create ghosts and snow. If you're running twinlead cable, twist it so it turns once every 18 inches or so. The twist will help keep out interference.

**Testing** When all is finished, turn your TV on and test each portion of your system. Is your VCR working properly? Can it receive all of the channels it's meant to receive? Is the picture crystal clear like it's supposed to be? If you're not sure what the picture should look like, hook the cable or antenna wire directly into the TV, bypassing all the other video gear.

**Debugging** Problems in installation can occur, but finding the fault doesn't have to be difficult. Take one step at a time. If your VCR isn't receiving a signal, go through your diagram to see if you can find out why. If the problem doesn't lie in your engineering, check the wiring. Substitute cables from paths you know are good. If you must, temporarily reroute or replace cabling to pinpoint the cause.

## 86 Checkout

After installation is complete, plug in the VCR. Note the polarization of the plug and socket, as shown in Fig. 3-8. Almost all electrical devices sold these days have polarized plugs—one prong is slightly larger than the other. They are designed to only fit one way into the wall socket.

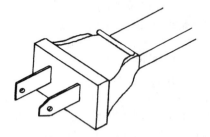

■ **3-8** *A polarized ac plug, now used on most electrical devices sold in the United States.*

If the electrical outlets in your home are not equipped with polarized outlets, be sure to orient the power cable from the VCR in the same direction as the power cable from the TV set, as shown in Fig. 3-9. By retaining the proper polarization for the two components, you eliminate the possibility of a ground loop. Aurally, a ground loop sounds like a low-volume, low-frequency hum. Visu-

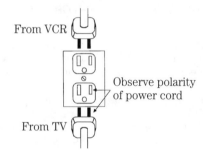

From VCR

Observe polarity
of power cord

From TV

■ **3-9** *Observe the polarity of the power cords when plugging in the VCR and TV. Reversing the polarity on one device and not the other can introduce electrical hum, which results in a dark band on the screen or in a raspy sound.*

ally, the loop appears as one or more dark bars that move slowly from top to bottom (or vice versa) of the screen.

Next, turn on the TV (if it isn't already), put a tape into the VCR, and tune the VCR to a broadcast channel. Record several minutes of the broadcast, then play it back. You should see a picture.

Check all of the front panel controls. Press the PAUSE, FAST FORWARD, and REWIND buttons on the deck to make sure everything works. If the VCR has additional features and special effects, consult the operator's manual and try them out.

## If something goes wrong

Okay. You've connected your VCR to your TV and did everything that you thought you should do, but it doesn't seem to work. Table 3-1 is a quick troubleshooting guide you can use to help correct any mistake. These problems assume a simple cause. For a diagnosis of more serious problems, see the troubleshooting flowcharts in chapter 9. Keep in mind that many video problems are not caused by the VCR but by the television set, antenna, cable, videotape, and installation. Always suspect a fault not related to the VCR first. Chapter 7 details non-VCR problems and how to solve them.

## Achieving optimum picture and sound

There are a number of ways you can improve the picture and sound to achieve maximum performance from the deck.

### Better picture

If your TV accepts baseband audio and video signals, use the AU-DIO and VIDEO OUT connectors on your VCR, not the RF terminal. The reason is that when transmitting the signal via the RF termi-

**■ Table 3-1 Basic troubleshooting guide—VCR installation.**

| Problem | Cause | Remedy |
|---|---|---|
| **Won't turn on** | Not plugged in | Plug into good socket |
| | Switched outlet | Switch on outlet |
| | Blown fuse | Replace fuse (usually internal) |
| **Tape won't load or eject** | Tape not inserted properly | Reorient tape and try again |
| | Power off | Turn power off |
| | Tape spilled inside deck | Remove tape from interior of VCR |
| **No sound or picture** | VCR not properly connected to TV | Check cabling |
| | Bad or incomplete connection | Check or replace cabling |
| | TV set not adjusted or wrong channel | Check controls on TV |
| | Tape blank or damaged | Insert known good tape |
| | Heads dirty | Clean heads |
| **Will not play tape** | Tape not inserted properly | Reorient tape and try again |
| | Power off | Turn power off |
| | Tape spilled inside deck | Remove tape from interior of VCR |
| | Condensation in deck | Wait 30 minutes to dry |
| **Picture is snowy** | TV set not adjusted or wrong channel | Check controls on TV |
| | Video heads dirty | Clean heads |
| **VCR does not respond to controls** | Tape not properly loaded | Reload tape properly |
| | Switches dirty or electrical problem | Turn unit off and on again |
| **Remote does not work** | Bad or missing batteries in remote | Install fresh batteries |
| | Infrared light path blocked | Unblock path (including glass) |

nal, the signal undergoes several unnecessary processing steps, and the sound and picture are degraded as a result. Whenever possible, use the AUDIO and VIDEO baseband terminals.

If you have a Super VHS deck and a television set equipped with Y/C terminals, be sure to use them. The Super VHS deck is only delivering a percentage of its rated resolution when connected to a TV without the separate Y/C terminals (for more information on Super VHS, see chapter 2).

You can enhance the picture quality by adding a signal processor, as discussed in the "Video accessories" section in this chapter.

## Better sound

By connecting your VCR into your hi-fi system, you greatly improve the quality of the sound. Use a good amplifier and set of speakers. If your VCR is mono, you can use a Y adapter to branch the one audio line from the VCR to the right and left channels of the hi-fi (see Fig. 3-10). Plug the VCR into the hi-fi terminals marked TAPE or AUX. Never use the PHONO input; the voltage output of the VCR is too high for the PHONO input, and you run the risk of damaging the stereo.

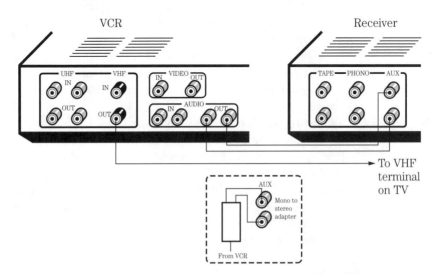

**89**

■ **3-10** How to hook up your VCR to a stereo receiver or hi-fi amplifier. Use a Y (mono-to-stereo) adapter when connecting a mono VCR to the receiver.

For optimum sound quality, keep these points in mind:

☐ Position the speakers so they are directed to the central listening point in the room. That point shouldn't be any closer than about 8 feet from the speakers. The spot where the sound from the two speakers meet is called the "sweet spot" (Fig. 3-11 on the next page) and is where the illusion of the stereo image is the greatest (assuming you have a stereo VCR).

☐ Position the right and left speakers evenly so they are the same distance away from the sweet spot. This arrangement increases the stereo image.

☐ Speakers almost always sound better when they are positioned on the long wall of the room. If the sound is too "boomy," the speaker placement in that particular room might be causing standing waves. Try a new location.

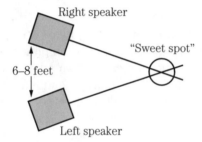

Right speaker

"Sweet spot"

6–8 feet

Left speaker

■ **3-11** *The "sweet spot" is the intersection of the sound waves from the right and left speakers.*

☐ Place speakers from 6 to 8 feet away from each other. Placing them closer together diminishes the stereo effect; farther apart creates a sonic "hole."

☐ Avoid placing the speakers in the corner of the room. This positioning diminishes smooth frequency response.

☐ Keep speakers off the floor, particularly if the floor is carpeted.

☐ It's best to position the speakers 2 to 3 feet from all walls. The closer a speaker is placed to a wall, the more bass it produces. This tip can't always be observed, but try to follow it as much as you can.

☐ If the stereo image is blurred or lacking depth, there might be excessive mid and high frequencies. Dampen these by placing a wall hanging or other soft absorbing material on the wall between the two speakers.

## Proper tape handling

One of the earliest debates to hit the video community was on how to store videotapes. Should they be stored stacked one on top of the other, or on their ends, or backs, or fronts, or what?

If you look at the construction of a videotape cassette and take a hint from the old time audiophiles, the answer is fairly clear. It's a generally accepted fact that audio recording tape shouldn't be stored flat (an open reel tape resting on its hub, for example). The strain caused by the reel and the weight of the tape itself can cause warping. For best results, the tape should be stored upright, so the weight of the reel and tape is distributed more evenly.

Videotapes use internal reels, too, so the same rules of thumb apply. When storing a videotape, place it upright (on edge) either on the spine or one of the ends. The spine is where you place the label that describes the contents of the tape, so if you want to choose easily your tapes from the collection in the bookcase, store the cassette on the top or bottom edge.

Technically, it's probably best to store the tape so the full reel is on the top (the weight of a full reel on the bottom might stretch the tape), but the videocassette then tends to be top heavy. A perfectly acceptable way to store tapes is so the empty reel is on the top. A ratchet mechanism inside the cassette, as shown in Fig. 3-12, prevents undo tape tension so the cassette won't be top heavy.

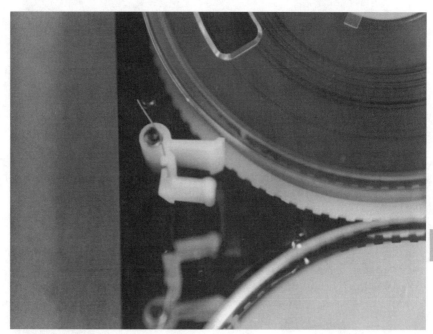

■ 3-12 *The ratchet and pawl mechanism in VHS and Beta cassettes prevents the tape from spilling off the spools. The pawl is released when the cassette is inserted into the VCR.*

By the way, there's no rule that says you must store videotapes in their cardboard or plastic dust jackets. Many seasoned videophiles throw out the tape covers because they get in the way. If you don't use covers, you should at least try to keep the tops of the tapes dust-free. If you live in a dusty area and don't like constant housekeeping, consider buying a videotape storage cabinet. Make sure it lets you store the tapes the proper way (most do, but some poorly designed cabinets overlook correct tape keeping).

## Tape life

Contrary to what you might have heard, there is no real need to mark the number of times you play your videotapes. A popular myth says that you can use a tape only a certain number of times. This myth can be true, of course, but that number is in the hundreds, and you'll likely never use a tape that many times.

Each time you play a tape, you scratch off a little bit of the oxide coating that stores the video and audio signals. At first, the amount isn't much, but after some time—after the tape has been played five or six dozen times—the results become visually obvious: flashing streaks of white begin to appear in the middle of the picture.

Should you throw the tape out? Not at all. You'll get at least 50 more plays out of the tape, and without the worry of damaging the deck. If, in fact, the tape gets so worn out that it dirties or clogs the video heads, it's a simple matter to clean them. When the tape becomes unviewable or repeatedly clogs the heads of your VCR, it's time to toss it.

## Damaged tapes

That takes care of normal tape wear. It's a different matter when the tape is physically damaged. A number of things can hurt a tape. By far, the most common are:

☐ Misthreaded tapes—where the tape jumps the tape guides inside the deck and gets tangled in the mechanism.

☐ Smashed tapes—where the door of the cassette closes on tape that hasn't been wound back onto the reels.

☐ Eaten tapes—where the take-up reel in the VCR doesn't keep up with the tape winding through the transport, and extra tape spills into the works.

Damage doesn't always mean the tape is unwatchable. If the hurt is minor, the tape will still pass through the VCR without harming the heads. There will be a noticeable amount of picture noise for the duration of the damage, however. Severe damage, where the tape is badly mangled, torn, or stretched, means that it's unusable and shouldn't be played.

Videotape can be spliced, but it is safe only when the splice is at the very beginning or end of the tape (the leader portions). That's why it's always a good idea to fully rewind a tape before ejecting it; damage can often be repaired at the leader portion without losing the entire tape. Details on how to splice videotape are in chapter 5.

For best results, keep your VCR in top working condition. An old, dusty VCR invites tape problems. For example, if the idler wheel gets old and cracked, the take-up reel will slip as the tape is played through the deck. That causes tape to spill—the VCR "eats" the tape, ruining it. This common problem can be avoided by proper maintenance, as described in full detail in chapter 5, "General cleaning and preventive maintenance."

# Video accessories

There was a time—not long ago and in a place not far away—when the amateur video enthusiast endured tapes with poor colors, fuzzy images, and rocking and rolling pictures. But that was the limit of the state of the art just five short years ago. Now, however, with more than one quarter of every American household with a VCR, add-on components—little black wonder boxes—have come along to clear up video's bad image.

But there's a catch. You've probably seen video accessories—image enhancers, color processors, RF switchers, sync stabilizers, and a slew of others—advertised nearly everywhere. To the uninitiated, their purpose isn't always clear. Here's a rundown of the popular types of video accessories. Note that many of the accessory functions are now available in one, all-purpose box.

### RF and baseband video-routing switchers

How would you solve this problem? You have two TVs—one in the living room and one in the bedroom. Attached to the TV in the living room is a VCR and a cable TV box. The set in the bedroom also ties in to the cable TV box, but it's connected in such a way that whatever you're watching from cable on the living room set also is on the bedroom TV. You're not wired up so you can watch the VCR in the bedroom.

You're thinking about buying a disc player, but you're unsure of how or where to hook it up. Actually, you'd like to rearrange your entire system so it could be more versatile and convenient. In fact, right now, your home video system is so terribly limiting, you find your time spent watching the tube unrelaxing.

Does this scenario sound familiar? If it does, you're in good company. Most video enthusiasts have too many options and no way to choose among them. Yet the solution is surprisingly simple: routing switchers. Available in two types—RF and video—routing switchers let you add full flexibility to your video gear with very little headache.

The initial choice of the type of switcher to get—either RF or baseband video, as detailed in chapter 2—largely depends upon your system and what you want to do with the signals. If, for example, you have a component TV and the rest of your video gear can provide baseband video outputs, you might do well with a video switcher. With such a setup, you'd be able to select between the TV tuner, VCR, disc player, satellite dish, or other video source easily.

Add a second component TV, however, and things get messy. To be used effectively, each component set must have its own tuner. Unless you wire up the second TV's tuner to the antenna or cable system separately, you'll be better off switching entirely to RF because you cannot combine video signals and RF signals in one cable or switcher. The bottom line is that video switchers are best for single-TV environments.

For most systems—or at least those that serve more than one television set—an RF switcher will be required. It provides the greatest amount of flexibility and is readily adaptable to a variety of equipment. With RF, you can send a number of signals at different frequencies through the system at once and let the TV select the proper one (remember, with video, you can only send one signal at a time). With RF, what you lose in quality, you gain in convenience.

However, you'll want a baseband video switcher if you have a TV/monitor that accepts direct video and audio inputs. Connect the outputs of your VCR, disc player, and other gear equipped with VIDEO OUT and AUDIO OUT jacks to the switcher. You can then "dial" in the source you want to view on your TV. Note that a number of the TV/monitor sets have their own built-in video switchers to select between two or three separate sources.

Common to all switchers, RF or video, is their *switching matrix*, or the function that determines their overall switching capability. Exactly what is a switching matrix? It is the maximum number of inputs by the maximum number of outputs the switcher can accommodate. For example, a switcher with a $5 \times 3$ matrix has five inputs and three outputs. That means that any of the five input signals can be selected at each of the three outputs.

As you recall, RF signals are used to broadcast programming over long distances, so they have an uncanny knack of sneaking into all sorts of places. The result is that when used in a home video system, RF signals tend to radiate and leak into places you don't want them to. That's why when working with RF, it is important to get a switcher that has good RF isolation. The higher the isolation rating of the switcher, the better it is in keeping signals from leaking and interfering with each other. A practical minimum is 40 dB of isolation, although a good switcher is rated at about 60 to 80 dB. As you might have guessed, the higher the number, the better the isolation, and the less chance you'll see interference.

94

Isolation becomes a critical factor when you send two or more signals on the same channel into the switcher. For instance, you might have two VCRs—both of which have channel-3 outputs—feeding into one switcher. On an inexpensive, low-isolation switcher, these two signals will surely cross and cause undue interference and ghosting.

Another major selection criterion is the amount of signal loss—called *insertion loss*—inherent in the switcher. A certain amount of loss in any switcher is natural: every time you take one signal and divide it a number of ways, you're going to make it weaker at each output.

The best video switchers include a built-in amplifier that maintains the proper signal level at each output, no matter how many times the signal is branched. This amplifier is vitally important because equipment that relies on video signals expects those signals to be at a precise level. Too much or too little—even a fraction of a volt—can mean problems. Insertion loss in baseband video switchers is typically very low.

With an RF switcher, you have far greater latitude when it comes to signal levels. The circuitry built into gear meant for RF reception can largely compensate for shortages and overages in signal level. But RF selectors have far greater insertion loss, so you can't carelessly pop an RF switcher in your system and expect the best. RF switchers that have two or three outputs typically have a 7 to 10 dB loss.

With signal loss, unlike isolation—where the higher the number, the better—you want a low number. Signal loss will have negligible effects if what you're feeding into your switcher is strong to begin with. A marginal signal (one that looks grainy or washed out), or a weak signal (one that looks "snowy") can be reduced so much by the switcher that it could make viewing difficult. Opt for a switcher with a low insertion loss if you are plagued with weak pictures.

For video switchers, you'll want to consider getting one with an audio switching capability built in. Several models can switch in stereo. These units provide connections for a video signal plus two separate audio signals for each source. If the unit doesn't have an audio capability, you'll need to buy yet another switcher for the audio portion of the signal.

## RF modulators

RF modulators combine a video signal with an audio signal and transmit them over a cable on either TV channel 3 or 4. The better models allow you to select which of the two channels you want; the less expensive models are factory preset.

## Image enhancer

Another major accessory is the image enhancer, which is used to sharpen or add detail artificially to a video picture. Image enhancers work much like the audio equalizers found in home stereo gear: they selectively boost the higher frequency portions of a video signal. The higher frequencies are the ones that provide sharpness and detail. The more accent given to the high-frequency signals, the more detail and sharpness there will be.

Keep in mind, however, that no image enhancer can give back what was never there in the first place. In addition, image enhancers can't put detail back into something that was poorly recorded. The high-frequency signals just won't be there to boost.

Image enhancers are best used when making the recording in the first place, particularly when copying a tape. Repeated recordings will wash all the detail out of a tape, so it's helpful to boost the sharpness before the image is rerecorded. The better image enhancers allow you to control sharpness as well as noise, which is naturally introduced in the enhancing process.

## Color processors

The color processor is another useful video accessory. As explained in chapter 2, "How VCRs work," color is transmitted in the television band in such a way that its hue and intensity can be artificially manipulated. To increase or decrease color, the level of the burst signal can be weakened or strengthened. To change color, the phase of the burst signal is changed.

But why change and manipulate colors? There are several reasons. One is that colors can shift, particularly when creating tapes with a color camera. Greens can become blues, reds can become greens and so forth. A color processor can be used to bring the colors back into line. Another reason is that old tapes can fade. Because a color processor can boost the intensity of color far beyond that originally recorded, it's possible to rejuvenate old or poorly recorded tapes with a single twist of a knob.

Old movies broadcast on television also are good contenders for color processors. Because of their age, they are often washed out. Adjust the *chroma gain* to increase the vibrancy of the colors; adjust the *chroma phase* to change the colors (the names of the controls differ from model to model, but most processors use this nomenclature).

Color processors also can be used to vary the overall brightness of the entire picture—color and all. You can bring the scene from normal brightness to pitch black. Separate boxes known as video faders do the same function and can be used to provide special effects to your home tapes. The best faders include an audio control as well, so both video and audio signals can be faded together.

## Video stabilizer

If you buy or rent commercially produced videotapes, there is one video accessory that you might not be able to get along without: the video stabilizer. Many of the companies that distribute prerecorded movies change the synchronization signals on the tape slightly to render the video signal uncopyable by a VCR. Most systems omit or invert some or all of the vertical synchronization signals that occur 60 times per second during video transmission.

Unfortunately, many TV sets—particularly the new, all electronic ones—can't handle this change in sync signals. The result is a picture that flips and rolls as if the vertical hold were maladjusted. On some sets, in fact, all that's needed is to adjust the vertical hold and the picture is stable.

But there's the rub. Most of the newer sets lack a vertical-hold control or have one inconveniently located on the back or even inside the chassis. The only alternative—apart from not watching copy-protected tapes—is to add a video stabilizer to your VCR.

Sync stabilizers simply put back that which was taken away (more or less). A knob on the front of the stabilizer allows you to fine tune the timing of the sync signals to get the best correction possible. You'll have to readjust the knob each time you watch a new tape, however.

A new form of anticopying process, called Macrovision, is used on nearly all of the newer movie releases. Macrovision disrupts the automatic gain control circuitry in VCRs, making VCR-to-VCR copying difficult. The result, as seen on the copied tape, is a picture that changes brightness with occasional bright flashes of light. The audio portion is unchanged.

Some television sets are susceptible to the Macrovision encoding scheme, and a number of companies have video stabilizers that attempt to correct for the offending Macrovision signals. The Macrovision process also can impair the picture when you are not copying. Chapter 7, "Troubleshooting and repairing non-VCR problems," addresses this topic.

## Stereo synthesizers/decoders

Stereo broadcasting is still fairly new. If you have a TV or VCR equipped with an MTS tuner (or a decoder), you can receive broadcasts that are transmitted in stereo. When the program you're watching isn't in stereo, you can still simulate the fuller sound with a stereo synthesizer. The synthesizer hooks up to your hi-fi and you can vary the amount of stereo effect.

Many synthesizers are combo units that also incorporate Dolby "surround" decoding, a stereo process used in movie theaters. With surround audio, you install an additional back (and sometimes front) speaker for added listening dimension. You must have a stereo VCR (hi-fi or linear) and play an encoded prerecorded tape (such as "Raiders of the Lost Ark"), or watch Dolby-encoded programming from a satellite or cable channel. Surround audio decoders also are available as stand-alone units without the stereo synthesizer circuitry.

## Replacement remotes

The remote control that came with your VCR isn't the only one you can use with it. Depending on the model, you can purchase third-party remote controls as a replacement for the remote control that came with your VCR. You can use the additional remote control if the original is lost or damaged. Or, you can use the additional remote as a backup or spare. You can even play "dueling remotes" where you change to your favorite channel using your remote, and the person watching with you changes using another remote!

Replacement remote controls come in all shapes and sizes. There are two general types of replacement remote controls on the market. Both types use infrared light to transmit commands to your VCR:

☐ So-called "universal" remotes come preprogrammed for use with your VCR. Before you can use the remote control, you need to identify the brand of VCR you are using. This information tells the remote control how to communicate to your VCR, using the proper sequence of light pulses. The disadvantage of universal remotes is that if your VCR is brand new, the current crop of remotes might not be able to talk with it. When in the market for a universal remote, be sure to buy the latest model, and be sure you can return it if it doesn't work with your deck.

☐ "Learning" remotes aren't preprogrammed for any VCR. Rather, they learn the specific codes the original remote control uses to command your VCR. Setting up a learning remote takes a bit longer, because you must program each control button individually—PLAY, RECORD, STOP, and so forth. The main advantage of the learning remote: unless you have an odd-ball VCR, you're assured of being able to use it with your deck. The main disadvantage of the learning remote is that you need the original remote to program it. If the original remote is lost or damaged, you won't be able to program a learning remote.

Not all third-party remote controls offer the same features. Basic models offer controls for just the basic features of your VCR, and little else. More expensive remotes add additional features that your VCR might have, like special effects modes. Most third-party remotes have no provision for allowing you to use them to set the clock on your VCR, or program your VCR to record programs while you are away. You need your VCR's original remote control for this.

If your VCR is not in the same room as your TV, you can purchase a "UHF" remote control. Rather than use an infrared light beam, UHF remotes use radio waves to transmit instructions to the VCR. The advantage of the UHF remote is that it works through walls; infrared remotes do not, and generally require you to be in the same room as the VCR.

Add-on UHF remote controls consist of two parts: the remote control itself, and the UHF receiver. The receiver is plugged into the wall socket for power, and is located near the infrared sensor on the front of your VCR. It acts as a type of relay station: press a button on the remote control, and the receiver captures the UHF signal. It then sends the proper infrared command signals to your VCR.

## Turbocharging your VCR

Your new VCR gleams with radiant beauty; it's a prized possession worthy of your adulation. Despite slight objections from your significant other, you bought the best—ah, most expensive—model you could find, determined to get the ultimate picture and sound money could buy. Alas, despite your sizable cash outlay, after bringing the VCR home and connecting it to your TV, the disappointment in the less-than-sterling video quality shows in your face. So does the gaping hole in your checkbook and the stern look of your better half.

Relax, there's no cause to be banished to the dog house—yet. Just like the family car, which runs smoother and gets better mileage after a few simple mechanical adjustments, the reliability and performance of your VCR can be greatly enhanced by making a few simple modifications in the way things are wired together. Instead of spark plugs and ignition timing, as a VCR "driver" you're concerned with such ingredients as cables, connectors, video levels, and signal-processing equipment.

Although it's possible to go overboard, consumed by an obsession to make constant adjustments and alterations to your video system, you'll be amazed how much better the sound and picture from your VCR can be improved just by paying attention to seemingly minor details. With a little effort, and only a marginal drain on your pocketbook, you can turbocharge your VCR to squeeze every last bit of performance from it possible.

## VCR hospitality

Be sure your VCR is in top operating condition before attempting the turbocharging process. If your VCR is not new, you might need to clean it to remove dust, dirt, and other contaminants. Clean the heads, as detailed in chapter 5, using a commercial head cleaning tape, or do the job manually with a sponge-tipped applicator and bottle of cleaner. Chapter 5 also details other regular maintenance like routine cleaning and lubrication you can perform on your VCR to keep it in top-notch shape.

VCRs also appreciate a clean environment. Performance is always better when you regularly clean the surroundings of the VCR. Airborne dust inevitably collects on the electrical contacts inside and outside VCRs and impairs signal quality. Every few days or so, dust on and around the VCR with an antistatic cleaning cloth. When not in use, cover your VCR to protect it from airborne contaminants.

Home VCRs are not temperamental, but they do have a limited number of serious dislikes. Shield your VCR from:

☐ heater/air conditioning vents
☐ windows
☐ magnetic fields (like loudspeakers)
☐ old tapes that are losing their magnetic coating
☐ close quarters with no ventilation
☐ liquids

While you're improving the lot of your VCR, check the proper operation of your TV, antenna, and cable system. Unless you're an experienced technician, leave TV servicing to the pros: the high voltages inside a television set can be lethal (the 117 ac volts in a

VCR that's plugged in is far less than the 20,000-plus volts lurking in a TV; and that voltage is there even when the set is unplugged!). Have the TV repair person degauss the TV screen to remove trace magnetism. This 20-second chore can go a long way toward improving the picture.

Shortcomings of your cable service should be checked by the cable company. Most cable companies are lax to repair "minor" inconveniences like a snowy picture or cross-channel interference, but if you and your neighbors complain often and loud enough, you should see results sooner or later.

## Cables: conduits for electrons

Unless your TV has its own built-in videocassette recorder, the connections between the VCR and the other components in your video system are by cables. Apart from the proper operation of your VCR, TV, and other major components, the most vital element to good picture and sound is the cables that connect everything together. You can expect only marginal performance from your VCR if your video system is riddled with old, worn, or incorrect cables.

As explained in this chapter, cables for home video come in two basic flavors: coaxial and shielded. In most home video setups, coaxial cable is used to transmit RF signals, those that are carried over TV channels. Shielded cable is used for baseband audio and video, the type of signals available at the AUDIO OUT and VIDEO OUT jacks of your VCR. Though the fundamental construction of the two types of cables is similar, coax is much thicker and bulkier than shielded. Shielded cable also is often used in hi-fi systems to connect turntables, compact disc players, and tape decks to an amplifier, so it's probably the most familiar to many.

This discussion brings up an important point: if you're using shielded cable designed for hi-fi use in your video den, get rid of it. Why? All electrical circuits, including cables, have a certain impedance (recall that the technical definition of impedance is the opposition an alternating current encounters when it passes through cable or a circuit). Impedance is measured in ohms, and when the signal encounters different impedances throughout a circuit, its characteristics can be changed.

Your VCR is designed with 75 $\Omega$ input and output circuits (exception: microphone and headphone jacks have different impedances to match the equipment they connect to). Shielded hi-fi cable is not typically rated at 75 $\Omega$ impedance, so using it causes an imbalance in your video system. Using cables with the wrong impedance can cause serious loss of signal, ghosts, and interference. This fact is true for both audio and video.

Instead, use shielded cable made especially for video; it has an impedance of 75 $\Omega$. It costs more than hi-fi shielded cable, but the improvement in sound and picture quality is well worth it. Perhaps an even better approach is to use 75 $\Omega$ coaxial cable for baseband audio and video. If your TV has separate baseband audio and video inputs, for example, feel free to stretch a length of coax between the AUDIO OUT and VIDEO OUT jacks of your VCR to your television set.

Coaxial cable offers better rejection of interference (caused by alien signals traveling through the air that penetrate the cable and get mixed in with the ones you want), but it costs more than shielded cable and is generally harder to work with.

While you are replacing hi-fi cables with specially manufactured video shielded cables or coax, be on the lookout for unnecessary RF connections between the various components in your video system. Yes, RF is a convenient method of transferring the signal from one VCR to another or from a VCR to a TV but the extra processing steps required to transform video and audio signals into a radio-frequency television channel can corrupt the quality of the sound and picture. If the components in your video system have baseband audio and video jacks, use them. In most TV dens, the only RF connection should be at the antenna input or cable box or possibly at the antenna terminals of the TV.

Besides impedance, electrical cables also are affected by two other peculiarities: resistance and capacitance. Resistance is the electrical opposition encountered by a direct current. Like impedance, resistance also is measured in ohms. While carrying video and audio signals to and from your VCR, the resistance of a cable roughly determines how much of that signal reaches its destination. A cable with a high resistance acts to limit the signal, so the signal is effectively reduced by the time it finishes its journey. Conversely, a cable with a low resistance does little to impede the easy flow of electrons traveling between your VCR, TV, cable box, or other component. Signal strength is virtually untouched.

The resistance of cables used to carry RF and baseband video signals is not as critical as it is in some other electronics applications, for example, the wires connecting a high-powered audio amplifier to a set of loudspeakers. But we're interested in making as many small improvements as possible with an eye on achieving a greater overall increase in performance.

102

You might want to consider using video cables especially engineered for low resistance. These cables, such as those by Monster, are available at most high-end stereo and video shops. They work by being physically larger than most video cables, using many individual strands of wire inside. Electrons passing through the cable have a tendency to travel only on the outside of the individual wire strands. This effect is known as *skin effect* and is more apparent with high-frequency signals (such as the video from your VCR) than with low-frequency ones. By using more strands of wire, the electrons passing through the cable have more surface area to travel over. That, at least as the marketing talk goes, equates to lower overall cable resistance.

If resistance is the ability to pass a signal through an electrical circuit or wire, then capacitance is the ability to store that signal. All electronic circuits—and of course cables—exhibit capacitance. You guessed it—too much capacitance can ruin the signal. Imagine a signal that alternates between two levels. Ideally, the signal is composed only of straight horizontal and vertical lines, but add the effects of capacitance and the signal becomes "lazy"—the straight lines become rounded. If the capacitance is bad enough, the characteristics of the signal change so much that it doesn't even look like the original.

Most well made cables for video—either coax or 75 $\Omega$ shielded—exhibit little capacitance, on the order of only a few picofarads (a common measurement of capacitance) per foot. Rather, capacitance increases with poor or dirty connections (see the following section) or when the cable becomes old or deformed.

Coaxial cable used outdoors to connect your antenna to your TV or VCR should be replaced every five to seven years, more if the climate in your area varies between freezing winters and boiling summers. Again, video cables can easily become deformed by walking over them, so avoid routing cables—especially coax—under carpeting or mats where you are likely to step on them.

## Cable and connector gremlins

In addition to increased capacitance by wear or damage, cables and their connectors are subject to other signal-impairing gremlins. Correcting and avoiding these complications goes a long way toward improving the sound and picture of your VCR.

As mentioned, avoid coiling the excess length of cable in a tight loop. The loop forms an inductor that invites interference. In fact, when possible, use the shortest cable possible for each connection. Commercially made cables typically come in lengths of 1, 2, and 3 feet; buy an assortment and use the length required to do the job—no more, and no less.

You also might want to make your RF cables, thereby cutting each one to exact length. See appendix D for more details. Follow the same basic procedures when making baseband cables with RCA phono plugs. Phono plugs typically require soldering, although the crimp-on variety also are available through some electronics outlets.

Keep signal cables away from ac cords, and never coil the two together. If you can't help keep signal cables and ac cords separate, at least strive to cross the ac cords only at right angles to the signal cables. This precaution minimizes hum caused by ac induction. Similarly, keep signal cables away from magnetic fields. Route them away from loudspeakers and the back of the TV (where there is a slight but detectable magnetic field when the TV is on).

Video cables use untinned copper wire to carry the signal from place to place. Copper readily oxidizes, and the oxidation acts as a virtual shield to electrons. Whenever possible, use only sealed, molded cables. The all-in-one construction of the cable prevents oxygen from reaching the wire strands inside. Some high-end hookup cables, such as Monster cable, are promoted as "oxygen free" to prevent signal-starving oxidation.

If you make your own cables, you can help prevent oxidation by sealing the cable ends and connector. Apply a light coat of acetone-free silicone sealant (such as RTV) to the joint between cable and connector. Don't get any of the sealant on the contacting points of the connector. Let dry and plug the connectors in place.

Another method is to apply heat-shrink tubing over the joint. Before attaching the connector, slip a ¾-inch length of heat-shrink tubing over the end of the cable. Secure the connector. Next, slide the tubing over the back end of the connector. Apply a small flame from a lighter or match to the tubing until it shrinks to fit. Avoid excessive heat or you'll melt the tubing. You can buy heat-shrink tubing at Radio Shack and most electronics outlet stores. It comes in various diameters, lengths, and colors.

F connectors leave the center conductor of the coaxial cable bare. The conductor makes contact with a sleeve inside the mating connector of your VCR or TV. Because this center conductor is exposed—and is made of copper no less—it is subject to the same oxidation that befalls all other copper-based cables. Check all F connections in your video system and look for oxidized center conductors. Unoxidized copper looks bright and shiny; oxidization darkens and dulls the metal.

Oxidation is nearly impossible to remove without harming the conductor; a better approach is to remake the connection. Cut off the existing F connector and trim away the insulation and outer braid; a complete step-by-step description of attaching an F connector to a coaxial cable is provided in appendix D. Before adding the new connector, inspect the conductor to make sure the oxidation has not "crept" up into the cable and affected more than just the exposed tip.

## Connector quality

Not all connectors are created equally. The connectors used on most video cables are mass produced using copper or tin and then plated with nickel (sometimes chromium). The plating resists corrosion and oxidation. Nickel is only a marginal conductor of electricity, but its low cost and ease of manufacture make it the best all-around choice for consumer video.

Better quality connectors use gold plating. Gold is perhaps the best overall conductor of electricity (silver is the best because it passes current more readily, but it tarnishes). Gold never corrodes, but alas, it's agonizingly expensive. Gold-plated connectors cost between two and four times as much as their nickel-plated cousins. You can purchase cables with gold-plated connectors already attached, or buy your gold-plated connectors for use on your own custom-made cables. The cables and connectors are available at Radio Shack and elsewhere in both the RCA phono and F-connector varieties.

Does gold plating really improve the sound and picture quality? There's no doubt that it plays a significant role in reducing cable resistance and capacitance (it does not affect impedance). Although you might need an oscilloscope to see a marked difference, the signal reaching the TV from your VCR is less affected when using gold-plated connectors than when using nickel-plated ones. The better your TV and VCR (especially if they too have gold-plated jacks), the more visual and aural proof you'll have.

One type of gold-plated connector probably won't do you much good. That's the gold-plated F connector. The gold plating is on the barrel of the F connector, which is the ground connection for the video signal. The center conductor of the coax cable comprises the actual signal-carrying wire, and it's not gold-plated.

Whether you use gold- or nickel-plated connectors, you can help decrease cable resistance and capacitance by thoroughly and periodically cleaning the connectors in your video system and by using a "contact enhancer" such as Sumiko's Tweek. Clean the connectors on your cables, TV, VCR, and other gear with a cotton swab soaked in denatured alcohol (undiluted if possible). Wait a few minutes for the alcohol to evaporate, then apply a thin coat of contact enhancer. Make it a habit to repeat the process every two to four months.

Never attempt to clean cable connectors with a file or emery board, and never (got that—never!!) use a cleaner/lubricant such as WD40. The WD40 spray is water insoluble and electrically nonconductive, so using it can render your entire video system inoperable.

## Which coax?

Coaxial cable is available in a number of different impedances. The proper impedance for video is 75 Ω. Both the RG-6 and RG-59 varieties of coax cable—available at almost any TV, video, or electronics store—is rated at 75 Ω. RG-6 is heavier and uses a larger center conductor than RG-59 and is considered the all-around better cable when using long cable lengths. The reason is that there is a lower signal loss per 100 feet of RG-6 than with RG-59.

However, RG-6 has a higher capacitance per foot than RG-59 (see Table 3-2). So if you're using short lengths of cable—less than 4 or 5 feet—the better choice is really RG-59. The best news is that RG-59 is less expensive and generally easier to work with. Note that most F connectors are made for RG-59. If you are using RG-6, be sure to get the F connectors with the larger crimp barrel.

Cable loss is expressed in decibels (dB) at different TV broadcasting frequencies. Every 3 dB drop doubles the signal loss. Note that signal loss increases at higher frequencies. Capacitance is expressed in picofarads (pF) and is cumulative for the length of the

■ Table 3-2 Common 75-ohm coax cables for video (specifications typical).

| Type | Outside diameter | Loss per 100 feet | Capacitance/foot | Center cond. |
|---|---|---|---|---|
| RG-59 | 0.242 inch | 50 MHz, 1.8 dB<br>100 MHz, 2.8 dB<br>500 MHz, 7.5 dB | 25.5 pF | 22 gauge |
| RG-6 | 0.266 inch | 50 MHz, 1.2 dB<br>100 MHz, 1.8 MHz<br>500 MHz, 5.1 dB | 18.6 pF | 18 gauge |

cable. That is, if cable capacitance is 15.5 pF per foot, a 3-foot length has a total capacitance of 46.5 pF. The higher the capacitance, the more the signal is degraded.

# Setting up a home theatre

One of the hottest buzzwords these days is *home theatre*. A home theatre is your video den made out to be as much like the local movie theatre as possible. It generally consists of:

☐ A large screen TV (27 inch or larger)
☐ A high-quality VCR and/or laser disc player
☐ An audio/video receiver (preferably with Dolby Surround Sound capabilities)
☐ Stereo speakers strategically placed on either side of your TV
☐ Other speakers positioned around the room, as desired

The concept of the home theatre is to bring home as much of the aural and visual excitement you feel when attending a quality movie theatre. The picture is large and bright, and the sound fills the room. You don't need to spend lots of money on a home theatre setup, but some people have been known to pay tens of thousands of dollars to bring movie experience into their dens. Fortunately, you get to decide on the amount of home theatre you want, and what you can afford.

If you're thinking about installing a home theatre in your house or upgrading your existing video setup to home theatre status, you need to think about each component as part of the whole. Though you might not be able to afford all the components at once, it pays to plan your system ahead of time and purchase the right equipment in stages.

The home theatre system combines audio and video, which means you have to spend more time wiring everything together. Unless you use the audio amplifiers in your TV (not really recommended for a bona fide home theatre setup), you will need to connect the audio and video outputs of your VCR to an A/V (audio/video) receiver. The output of the receiver connects to your TV. If you have more than one video source—you have a direct broadcast satellite dish for example, or a laser disc player—these are connected to the A/V receiver as well. Buttons on the front of the A/V receiver let you select the program source you wish to watch.

For listening to the sound, the A/V receiver is connected to two or more speakers. A pair of front speakers provide left and right stereophonic channels. If your A/V receiver is equipped with a

Dolby Pro-logic Surround feature (highly recommended), you will use the speaker in your TV, or an added "center" speaker, to fill in the sound between the right and left stereo speakers. You also will need to set up smaller speakers in the back of the room. These small speakers provide ambiance (or "surround"), and help to make you feel like you're immersed in the action you see on the screen.

# Tools and supplies
# for VCR maintenance

4

THE TOOLS AND SUPPLIES NECESSARY FOR PROPER VCR maintenance are not expensive, and you don't need an extensive set of tools. Other than common household tools, you need only a few specialty items to maintain your VCR in good operating condition and to diagnose minor problems. This chapter details these tools and supplies and how to use them.

## Workspace area

Regular VCR upkeep does not require removing the machine from its comfortable nest in your video system. As long as you dust the cabinet of the deck regularly and provide adequate ventilation for keeping the power supply cool, you need only remove the VCR for repair or a preventive maintenance interval.

For minor repair and preventive maintenance, you need a work area that's reasonably free from dust, is well lit, and is comfortable for you. Avoid taking your VCR out to the garage, where there is a greater chance for dust and airborne oil to contaminate its inner workings. A kitchen table or inside work area is ideal.

Before going to work on the deck, lay out a small piece of clean carpeting or heavy fabric over the work table to protect the table as well as the machine from scratches and dents. Collect all the tools you'll be using and have them on hand, preferably in a toolbox. Special tools and supplies can be stored in an inexpensive fishing tackle box. Tackle boxes have lots of small compartments for placing the screws and other parts that you remove. If you don't use a box, borrow a cup or bowl from the kitchen temporarily to store parts you remove from the VCR.

For best results, your workspace should be an area where the VCR will not be disturbed if you have to leave it for several hours or several days. The work table also should be one that is off limits or inaccessible to young children, or at least an area that can be easily supervised. Some chemicals used to clean VCRs are highly toxic, so you should keep them out of reach of children.

More important, there is a risk of electric shock with the top of the VCR removed. Take every precaution to avoid injury and never leave the deck unattended where curious fingers can touch high-voltage wires.

## Basic tools

You'll need a screwdriver and a few other common tools to disassemble the deck. Most VCRs use Phillips screws to contain the cabinet, chassis, and internal components, so be sure you have a Phillips screwdriver handy. Some decks use flathead or hex screws, but these are the exception, not the rule. Determine which tools you need ahead of time and get the proper ones. Don't try to make do with the wrong tool. Using a small flathead screwdriver to loosen an Allen screw only strips the head of the screw.

If your screwdrivers are not already magnetized, purchase a screwdriver magnetizer at the hardware store. When magnetized, the screwdriver holds on to screws that you remove or re-install into the VCR. The magnetizer lets you magnetize and demagnetize your screwdrivers and other metal tools. Or, you can magnetize your screwdrivers by scraping the blade across the face of a large speaker magnet.

Be sure to save all the screws you remove; they are not as easily replaced as you think. Most VCRs are made in Japan and use hardware with Japanese metric threads, or they use special metal or plastic self-tapping screws. You can't easily find them at hardware stores and they can be expensive if purchased at specialty outlets.

A pair of pliers also is a handy tool to have to loosen or tighten nuts, grommets, and plastic standoffs. Tweezers or a pair of small long-nosed pliers help you grasp small parts—like screws that have fallen inside the deck! A pair of regular manicuring tweezers is fine, but try to get the type with the flat, blunt end. Tweezers with a pointed end aren't as useful.

## Volt-ohmmeter

A volt-ohmmeter is used to test voltage levels and the impedance of circuits. This moderately priced electronic tool is the basic requirement for intermediate VCR maintenance and repair and is necessary for anything beyond routine cleaning. If you don't already own a volt-ohmmeter you should seriously consider buying one. The cost is minimal considering the usefulness of the device.

There are many *VOMs* (volt-ohmmeters) on the market today. For work on VCRs, you don't want a cheap model and you don't need an expensive one. A meter of intermediate quality is sufficient and does the job admirably. The price for such a meter is between $30 and $75 (it tends to be on the low side of this range). Meters are available at Radio Shack and most electronics outlets. Shop around and compare features and prices.

### Digital or analog

There are two general types of VOMs available today: digital and analog. The difference is not that one meter is used on digital circuits and the other on analog circuits. Rather, digital meters use a numeric display not unlike a digital clock or watch. Analog VOMs use the older-fashioned—but still useful—mechanical movement with a needle that points to a set of graduated scales.

Digital VOMs used to cost a great deal more than the analog variety, but the price difference has evened out recently. Digital VOMs, such as the one shown in Fig. 4-1 on the next page, are fast becoming the standard; in fact, it's hard to find a decent analog meter anymore.

Analog VOMs are traditionally harder to use because you must select the type and range of voltage you are testing, find the proper scale on the meter face, and then estimate the voltage as the needle swings into action. Digital VOMs, on the other hand, display the voltage in clear numerals, and with a greater precision than most analog meters. Because of their increased popularity and ease of use, this chapter concentrates on digital VOMs exclusively.

### Automatic ranging

As with analog meters, some digital meters require you to select the range before it can make an accurate measurement. For example, if you are measuring the voltage of a 9 V transistor battery, you set the range to the setting closest to, but above, 9 V (with most meters it is the 20 or 50 V range). Autoranging meters don't require you to do this, so they are inherently easier to use. When you want to measure voltage, you set the meter to VOLTS (either ac or dc) and take the measurement. The meter displays the results in the readout panel.

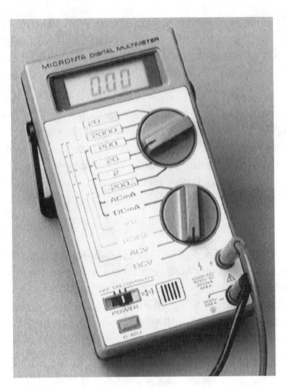

■ **4-1** *One of several dozen models of digital volt-ohmmeters.*

## Accuracy

A limited amount of the work you'll do with VCRs requires a meter that is super accurate. A VOM with average accuracy is more than enough. The accuracy of a meter is the minimum amount of error that can occur when making a specific measurement. For example, the meter might be accurate to 2000 V, ±0.8%. A 0.8% error at the types of voltages used in VCRs—typically 6 to 12 Vdc—is only 0.096 V.

Digital meters have another type of accuracy. The number of digits in the display determines the maximum resolution of the measurements. Most digital meters have 3½ digits, so it can display a value as small as 0.001 (the half digit is a "1" on the left side of the display). Anything less than that is not accurately represented; then again, there's little cause for accuracy higher than this when working on a VCR.

## Functions

Digital VOMs vary in the number and type of functions they provide. At the very least, all standard VOMs let you measure ac volts, dc volts, milliamperes, and ohms. Some also test capacitance and opens or shorts in discrete components like diodes and transistors.

For most purposes, these additional functions are not necessary, and you need not spend the extra money on a meter that includes them. To make effective measurements, you need to take diodes and transistors out of the circuit to test them accurately. The design of the latest VCRs makes this difficult and inadvisable, even for a seasoned repair technician. (Frequently, component failures are repaired by swapping the entire circuit board, not replacing components.)

The maximum range of the meters when measuring volts, milliamperes, and resistance also varies. For most applications including VCR troubleshooting, the following maximum ratings are more than adequate:

| | |
|---|---|
| dc V | 1000 V |
| ac V | 500 V |
| dc current | 200 mA (milliamperes) |
| resistance | 2 MΩ (megohms) |

## Meter supplies

Most meters come with a pair of test leads—one black and one red—each equipped with a needle-like metal probe. The quality of the test leads is usually minimal, so you might want to purchase a better set. The type with coiled leads are handy because they stretch out to several feet yet recoil to a manageable length when not in use.

Standard leads are fine for most routine testing, but some measurements might require the use of a clip lead. These have a spring-loaded clip on the end; you can clip the lead in place so your hands are free to do other things. The clips are insulated to prevent short circuits, and you can get clips that attach onto regular test leads.

## Meter safety and use

Most applications of the meter involve testing low voltage and resistance, both of which are relatively harmless to humans. Sometimes, however, you might need to test high voltages—like the input to a power supply—and careless use of the meter can cause serious bodily harm. Even when you're not actively testing a high-voltage circuit, it could be exposed when you remove the cover of the VCR.

If the deck is plugged in while the cover is off (which it will be if you're testing for proper operation) and you're not measuring voltages at the power supply, cover the power supply terminals, if exposed, with a piece of cardboard or insulating plastic.

Proper procedure for meter use involves setting the meter beside the unit under test, making sure it is close enough so the leads reach the test points inside the deck. Plug in the leads and test the meter operation by first selecting the resistance function setting (use the smallest scale if the meter is not autoranging). Touch the leads together, and the meter should read 0 Ω.

If the meter does not respond, check the leads and internal battery and try again. If the display does not read 0 Ω, double check the range and function settings and adjust the meter to read 0 Ω (not all digital meters have a zero adjust, but most analog meters do).

Once the meter has checked out, select the desired function and range, and apply the leads to the VCR circuits. Usually, the black lead is connected to ground, and the red lead is used at the various test points in the VCR.

Never blindly poke around the inside of a VCR in an attempt to get some type of reading. Apply the test leads only to those portions of the deck that you are familiar with—switch contacts, power contacts, and so forth. If you have a schematic diagram for the VCR, refer to it for the location of the test points.

One safe way to use the meter is to attach a clip on the ground lead and connect it to the chassis or circuit ground. Use one hand to apply the red lead to the various test points, and stick the other hand safely in your pocket. With one hand "out of commission," you are less likely to receive a shock if you aren't watching what you're doing.

## Logic probe

Meters are typically used for measuring analog signals. Logic probes test for the presence or absence of the low-voltage dc signals (bits) that represent digital data. These bits, or 0s and 1s, are usually electrically defined as 0 V and 5 V, respectively (although the actual voltages of the 0 and 1 bits depend entirely on the circuit). You can use a meter to test a logic circuit, but the results aren't always predictable. Further, many logic circuits change states quickly (pulse) and meters cannot track these voltage switches fast enough.

Logic probes, such as the model in Fig. 4-2, are designed to give a visual and (sometimes) aural signal of the logic state of a particular circuit line. One LED on the probe lights up if the logic is 0 (or low), another LED lights up if the logic is 1 (or high). Most probes have a built-in buzzer, which has a different tone for the two logic levels. That way, you don't need to keep glancing at the probe to see the logic level.

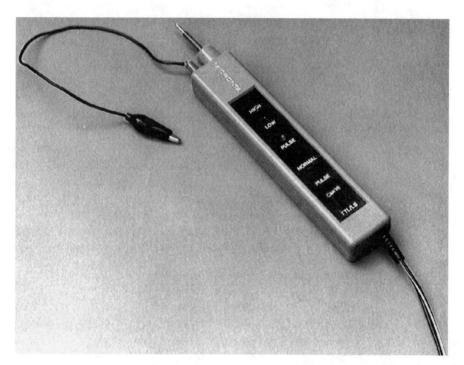

■ **4-2** *A logic probe. This one is switchable between transistor-transistor logic (TTL) and complementary metal-oxide semiconductor (CMOS) logic levels and can detect pulsing lines of up to 10 MHz.*

A third LED or tone might indicate a pulsing signal. A good logic probe can detect that a circuit line is pulsing at speeds of up to 10 MHz, which is more than fast enough for VCR applications. The minimum detectable pulse width (the time the pulse remains at one level) is 50 nanoseconds, again more than sufficient for testing videocassette recorders.

Although logic probes might sound complex, they really are simple devices, and their cost reflects this. You can buy a reasonably good logic probe for under $20. Most probes are not battery operated; rather, they get operating voltage from the circuit under test.

Unless you plan in-depth troubleshooting and repair of your VCR, a logic probe is not as important as a VOM. A probe is a handy tool to have should the need ever arise, but routine maintenance does not require it.

## Using a logic probe

The same safety precautions apply when using a logic probe as they do when using a meter. When the cover of the deck is removed, potentially dangerous high voltages might be exposed. If you are working close to these voltages, cover them to prevent accidental shock. Logic probes cannot operate with voltages exceeding about 15 Vdc, so if you are unsure of the voltage level of a particular circuit, test it with a meter first to be sure it is safe.

Successful use of the logic probe really requires you to have the circuit schematic to refer to. It's nearly impossible to use the logic probe blindly on a circuit without knowing what you are testing. A single VCR circuit board can contain components for both digital and analog signal processing, and you must know exactly what each component does and how it is used. And because the probe receives its power from the circuit under test, you need to know where to pick off suitable power. You can easily damage the probe—and possibly the circuit under test—if you connect the power leads incorrectly.

To use the probe, connect the power leads of the probe to a voltage source on the board, clip the black ground wire to circuit ground, and touch the tip of the probe against a pin of an integrated circuit or the lead of the component. Figure 4-3 shows a probe testing the logic level at an IC pin.

■ **4-3** Using a logic probe. Use care to avoid shorting the pins of ICs and other components. Use probe clips (available at most electronics stores) to anchor the probe to the test point.

Please note that VCRs often use negative voltages with respect to ground. To the logic probe, there is little difference between connecting the power leads to a positive and ground rail, or connecting the power leads to a negative and ground rail. However, avoid connecting the power leads to a positive and negative rail, because the voltage differential might exceed the maximum supply voltage of the probe. For instance, connecting the lead between the +9 and –9 rails will feed 18 V to the probe, which is higher than its rated operating voltage.

# Logic pulser

A handy troubleshooting accessory when working with digital circuits is the logic pulser. This device puts out a timed pulse, letting you see the effect of the pulse on a digital circuit. Normally, you'd use the pulser with a logic probe or an oscilloscope (discussed in this chapter). The pulser is switchable between one pulse and continuous pulsing.

Most pulsers get their power from the circuit under test. It's important that you remember this. With digital circuits, its generally a bad idea to present an input signal to a device that's greater than the supply voltage for that device. In other words, if a chip is powered by 5 V and you give it a 12 V pulse, you'll probably ruin the chip. Some circuits work with split (+, –, and ground) power supplies, so be sure you connect the leads of the pulser to the correct power points.

Also be sure that you do not pulse a line that has an output but no input. Some integrated circuits are sensitive to unloaded pulses at their output stages, and improper application of the pulse can destroy the chip.

## Making your own pulser

You can make your own pulser out of a 555 timer IC. A suitable schematic is shown in Fig. 4-4 on the next page. The pulser can be constructed on a small piece of perforated board or universal circuit board. Power for the pulser should be derived from the circuit under test (see the subsequent discussion) or it can be powered by a battery that delivers less dc voltage than the circuit you are testing. The 555 chip operates over a wide range of voltages, from about 3 to 15 V, making it suitable for interfacing with a variety of digital circuits. A finished pulser is shown in Fig. 4-5 on the next page.

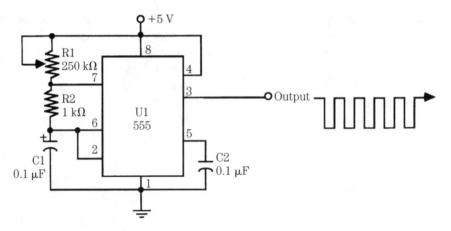

■ **4-4** *A homemade logic pulser. Vary potentiometer R1 to change the output rate.*

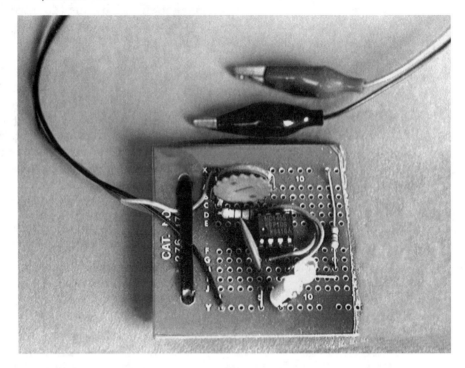

■ **4-5** *A completed 555-based logic pulser built on a universal project board. The unit is powered by the unit under test and can operate over a range of 3 to 15 Vdc.*

## Oscilloscope

An oscilloscope is a pricey tool—good ones start about $500—and only a small number of electronic hobbyists own one. For really serious work, however, an oscilloscope is a valuable tool, one that will save you hours of time and frustration. Things you can do with a scope include some things you can do with other test equipment, but oscilloscopes do it all in one box and generally with greater precision. Among the many applications of an oscilloscope, you can:

- [ ] Test dc or ac voltage levels
- [ ] Analyze the waveforms of digital and analog circuits
- [ ] Determine the operating frequency of digital, analog, and RF circuits
- [ ] Test logic levels
- [ ] Visually check the timing of a circuit to see if things are happening in the correct order and at the prescribed time intervals

The troubleshooting and maintenance procedures covered in this book do not require the use of an oscilloscope, but you'll probably need one if you want to delve deeper into VCR repair. Many service manuals and schematics for VCRs indicate the proper waveform at specific test points (the waveform is the visual representation of the electrical signal).

A basic, no-nonsense model is enough, but don't settle for the cheap, single-trace units. A dual-trace (two channel) scope with a 20 to 25 MHz maximum input frequency should do the job nicely. The two channels let you monitor two lines at once, so you can easily compare the input signal and output signal at the same time. You do not need a scope with storage or delayed sweep, although if your model has these features, you're sure to find a use for them sooner or later.

Scopes are not particularly easy to use; they have lots of dials and controls that set operation. Thoroughly familiarize yourself with the operation of your oscilloscope before using it. Knowing how to set the time-per-division knob is as important as knowing how to turn the scope on. As usual, exercise caution when using the scope with or near high voltages.

## Frequency meter

A frequency meter (or frequency counter) tests the operating frequency of a circuit, and like the oscilloscope, it is not an absolute requirement for intermediate-level VCR maintenance and troubleshooting. Most models, like the one shown in Fig. 4-6 on the next page, can be used on digital, analog, and RF circuits for a va-

■ **4-6** *A frequency counter switch selectable between 10 MHz, 50 MHz, and 500 MHz (a prescaler, not included in this model, is required for 500 MHz operation). This counter was built from a kit.*

riety of testing chores—from making sure the color crystal in the VCR is working properly to determining the radio frequency of the RF output of the deck. You need only a basic frequency meter—a $100 to $200 investment. You can save some money by building a frequency meter kit.

Frequency meters have an upward operating limit, but it's generally well within the region applicable to VCR applications. A frequency meter with a maximum range of up to 50 MHz is enough. One exception is testing the frequency of the RF output of the VCR, which can extend to 300 MHz and beyond. Higher-priced meters come with or have as options a prescaler—a device that extends the useful operating frequency to 500 MHz and higher.

## Infrared detector

VCRs use a number of infrared light sources for normal operation:

☐ Most VHS VCRs made after 1983 or 1984 use one or two infrared light-emitting diodes as light sources for the end-of-tape sensor. At the beginning and end of the tape is a clear portion (the leader). The infrared light passes through the leader and strikes the sensor. When the sensor detects the presence of light, it knows it has reached the beginning or end of the tape, so it stops. This action prevents tape damage during playback as well as during fast forward and rewind.

☐ Another infrared light source and sensor in the VCR is located under the take-up reel spindle. It detects when the spindle has stopped. If the spindle stops but the VCR is in the PLAY or RECORD mode, the deck automatically shuts down after 5 to 10 seconds to avoid excessive tape spillage.

☐ All but the least expensive VCRs come with a hand-held wireless remote control unit. The control works by transmitting a series of infrared pulses to the deck. When you push a particular button on the controller, a series of invisible pulses is emitted from the controller and received by a sensor in the VCR. The received light is then amplified and decoded, and the VCR acts on your command.

The infrared light sources are almost always *LED* (light-emitting diodes). In VCRs, they operate at about 780 to 904 nanometers of wavelength in the light spectrum, which is just outside normal human vision. Therefore, the "light" is actually invisible to your eyes, so there is no way you can visually check the operation of the various infrared LEDs in the deck.

An infrared detector can sense the presence of infrared light and provides you with a visual indication. The sensor is easy to build and is inexpensive. It can be used to detect the presence of all types of infrared light including that from laser diodes in compact disc players and videodisc players. That means you can use the sensor to troubleshoot and repair some of your other high-tech gear as well.

Keep in mind that you need not construct the sensor unless you are actively involved in the repair of VCRs or unless you firmly suspect that the LEDs in your deck or remote control unit have gone awry. Infrared LEDs are long-lasting solid-state devices with an average operating life of 10,000 or more hours, and they should outlast most of the mechanical components of the machine. Still, LEDs can break down after only a short time, or the circuit providing power to the LED could fail.

If you suspect that an electrical problem is preventing the LEDs from illuminating, you can test the circuit for power with a volt-ohmmeter (see the previous discussion). Position the leads on either side of the LED and take a reading (set the meter to dc V). Wireless remote control units also can be tested with an ordinary AM radio (see chapter 5 for details), although this method is not as foolproof because you are testing the operation of the circuits in the remote, not the actual light output.

Many older decks (built before 1982) use an incandescent lamp for the end-of-tape sensor. The lamp, which has a considerably shorter life span than an LED, might emit visible light, in which case you don't need the infrared detector. However, a number of the earlier decks used an infrared-emitting incandescent lamp. Beware of these; don't needlessly replace the lamp just because you don't see any light output. Use the infrared sensor to make sure that the lamp has indeed burned out.

## Circuit description

The infrared light sensor is a simple circuit that uses only three parts: an infrared phototransistor, an LED, and a resistor. The schematic is shown in Fig. 4-7. All the parts are commonly available at Radio Shack and most other electronic parts stores. You need some sort of battery supply to operate the circuit. We've built the circuit with four AA batteries enclosed in a self-contained battery compartment. The compartment also serves as the case for the sensor electronics.

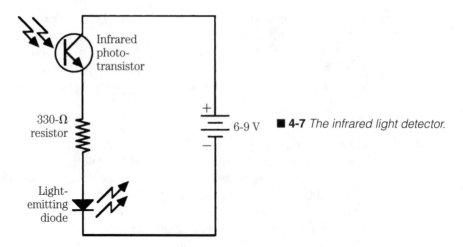

Infrared photo-transistor

330-Ω resistor

6-9 V

Light-emitting diode

■ **4-7** *The infrared light detector.*

Alternatively, you can use a 9 V transistor battery mounted in a simple clip holder. The resistor limits the current flowing through the LED, but the 330 Ω value selected is safe for use with supply voltages up to 9 V. If you use a higher supply voltage, increase the value of the resistor using Ohm's law. Failure to do this can cause the visible LED in the circuit to burn out.

You can determine the value of the resistor by taking the supply voltage and subtracting the voltage drop through the LED (usually 1.2 V). Divide the result by the current you want flowing through the LED (usually about 20 mA).

For example, with a supply voltage of 9 V, subtracting the drop in the LED makes 7.8 V. Divide that by 0.02 (20 mA), and you get 390. The exact value of resistor to use, therefore, is 390 Ω. Actually, there is a lot of slack built into this because LEDs can take higher currents without risk of damage. On a practical level, don't exceed 40 to 45 mA.

For best results, solder the components to a small perforated board, or use an 8-pin IC socket. The components fit into the contacts of the socket without soldering, but you might want to solder the leads in place anyway to prevent them from coming out.

Soldering into the socket is a little tricky because the plastic can melt. An easy way to do it is to fit the component leads into the contacts of the socket and briefly touch the leads with the tip of the soldering iron. Before the plastic of the socket melts, apply the solder. The electrical contact has already been made inside the socket by the contacts; the solder simply anchors the leads in place. Place the assembly in a box, like the battery holder shown in Fig. 4-8 on the next page, to prevent damage during storage and use. The LED is mounted on the other side of the socket so it can be seen when the sensor is pointed toward the LED under test.

Orientation of both the phototransistor and LED are critical. Installing them backwards will prevent the circuit from working. The phototransistor is marked with a tab or indentation so you can easily identify the emitter and collector. Refer to the specifications sheet that came with the transistor or consult a transistor guidebook. The plastic rim on the cathode side of the LED is almost always marked with a small indentation. You can install the resistor in any direction. Be sure to connect the power leads in the proper way, too—reversing the leads could destroy the phototransistor.

## Using the sensor

You can test the sensor by connecting the battery and pointing the phototransistor directly into a bright light (the sun and regular light bulbs all have infrared light content). The LED should glow. If it does not, check your work to make sure the components have been wired correctly (of course, be sure the battery is good; test it with your VOM).

Once you're sure the sensor works, you can test the LED in your VCR or remote control unit. You must get the phototransistor within 1 to 4 inches from the LED to pick up any light. Figure 4-9 on the next page shows the sensor being used to test the light output of the end-of-tape LEDs in a VCR.

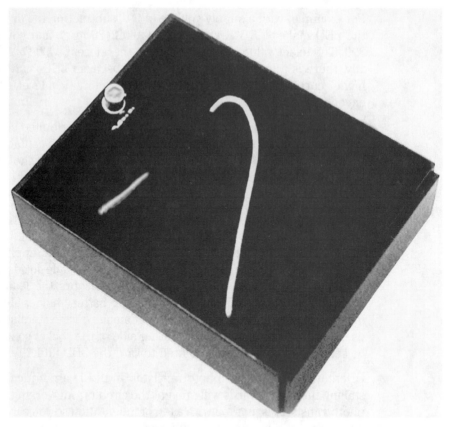

■ **4-8** *The finished infrared light detector built into the housing of a four AA-cell battery pack. The components are secured to an 8-pin IC socket.*

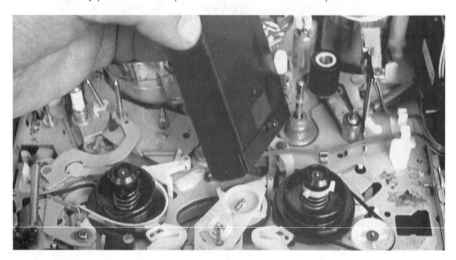

■ **4-9** *Using the infrared light detector to check the operation of the end-of-tape sensor and LED.*

## TV set or monitor

A small TV set or monitor is invaluable when troubleshooting and repairing a broken VCR. You can keep the VCR connected to the TV and test the deck as you go through the various troubleshooting procedures. It's much easier than hooking the deck back up to your living room set whenever you want to test something.

The TV need not be special, and unless you are troubleshooting problems with the color circuits, the set can be a black-and-white model. A surplus composite video monitor can be used with all VCRs that have a separate video output jack (99.99% do). Plug the VCR directly into the input of the monitor.

Keep in mind that most monitors lack a speaker, so if you want to hear sound, you need to provide a separate amplifier. A simple transistor battery-powered amplifier (the cost is under $15) can be connected to the AUDIO OUTPUT terminals of the VCR. Inexpensive, portable amplifiers use miniature ⅛-inch plugs. To connect one to your deck, use a phono-to-phono cable, or outfit one end with a phono to a ⅛-inch plug adapter.

If the amplifier doesn't have speakers, add one or two small ones. Almost any small speaker will do, as long as they are rated for use with the amplifier. Alternatively, if the amp has a headphone jack, you can use headphones to listen to the audio output. Use caution, however, because unexpected sound coming through the VCR can be deafening if the volume is turned up. When not needed, you should take the headphones off.

A pocket-size TV with a cathode-ray tube or liquid-crystal display also is ideal for VCR troubleshooting, maintenance, and repair. The TV has a built-in speaker (or earphone jack) and connections for attaching it directly to the VCR. Some pocket TVs have color screens, so you can test the color output of the deck.

## Assorted supplies

Unlike the family automobile, the family VCR requires little in the way of oiling and lubricating—if at all. Some cleaning and lubrication supplies might be necessary, however, and any well-equipped maintenance kit should have a little of both.

## Spray cleaner

The exterior cabinet of the deck can be cleaned with a damp sponge. If dirt and grime are a problem, use a mild spray household cleaner such as Fantastik or 409. Apply the spray to the sponge or cloth, not directly onto the cabinet. Excess can run inside and possibly cause damage.

## Cleaner/degreaser

The latest cleaner/degreasers on the market are "ozone friendly," meaning they use a solvent base that contains no hydrofluorocarbons (the stuff that eats away at the ozone layer). Although the latest cleaning supplies are more environmentally conscious, there's still plenty of cleaners out there that use the old fluorocarbon mixture. The mixture was most often Freon, a tradename for a popular brand of fluorocarbon solvent. In addition to being a good solvent for stuff like dirt and grease, Freon also is used (though much less now) as a coolant in air conditioners. Unlike most petroleum-based solvents, Freon doesn't melt plastics, and the type of Freon that you can readily buy is nontoxic and nonflammable. It does have a distinctive odor, but it is harmless.

Freon, or a Freon-substitute, by itself can be used as a basic degreaser and cleaner. You can use it to remove things like grime and dirt in hard-to-reach places. Freon is available at almost any industrial supply house.

The nonpetroleum solvent is often mixed with alcohol to make a more potent cleaner. Use caution with this mixture: it is flammable and toxic. The solvent and alcohol mixture can be used as a general cleaner, a degreaser, even as a video head cleaner. Chapter 5 shows you how to use the mixture to clean video heads and other delicate components.

The solvent/alcohol mixture is available as an all-purpose cleaner/degreaser that you can buy in a spray can. The cleaner is available at Radio Shack and most electronics supplies stores. The cleaners leave no residue, so you can spray it on, and it will dry with no trace.

The cleaner/degreaser can be used on all the internal components of the VCR, including the printed circuit board, loading and threading mechanisms, and record/playback heads. Always remember to turn the deck off and let it cool down before spraying. Otherwise, you run the risk of a short circuit. Worse, spraying the

126

cold liquid on warm parts can crack some components. This danger is especially present in rotary video and VHS hi-fi audio heads. Do not use a spray cleaner on these components if the deck has been recently operated.

## Grease and oil

Though VCRs have numerous mechanical components, most do not need any special lubrication. The reason: the parts are either impregnated with a lubricant (usually a high-viscosity oil) or are made with a material like Teflon that does not require lubrication. Some lubrication might be called for, however, and is especially recommended if the deck has been used in adverse environments (a portable VCR or camcorder would fall into this category) or is more than two or three years old.

Any light machine oil can be used for the components that need oiling. The oil should not have antirust ingredients. Good candidates are 3-in-1 oil or almost any oil designed for sewing machines. Another good oil is the type designed for musical instruments. This high-grade oil comes in a handy applicator bottle. Some bottles have a syringe-type needle for applying the oil in hard-to-reach places.

The best oil to use is the type designed for small machined parts. This oil, which is packed in a small bottle with a syringe applicator, has special penetrating lubricants that ordinary oils lack. You can buy this oil at most industrial supply houses and some camera and electronics stores.

If the oil you have doesn't come with a convenient syringe-type applicator, buy a set of disposable hypodermic needles at the drugstore. The needles are usually available behind the counter and can be purchased by adults without a doctor's prescription. The cost is under 50 cents per hypodermic.

To suck the oil up into the hypodermic, open up the can or jar of oil and dip the needle into it. Pull back on the plunger slowly; the oil will sluggishly seep into the hypodermic. To use the syringe, point the end of the needle directly on the spot you want to oil, and push gently on the handle. Apply only a very small amount of oil. Exercise care when handling the syringes and always keep the protective caps in place when not in use.

The "grease" should be a high-quality industrial lubricant, such as Lubriplate, or a nonpetroleum silicone product. There are a variety of lubricants from which to choose. A light-grade lubricant, such as that used in 35 mm cameras, is suitable. You can get it from most industrial supply stores and camera repair shops. A small tube goes a long way.

Refer to chapter 5 for more details on common VCR components that require oiling and lubrication. Note that some mechanical components do not require oil or lubrication and in fact can be harmed by them. Motors fall into this category. It's a safe bet that if there is no sign that oil or lubricant has ever been present on a mechanical part, it does not need it!

## Miscellaneous cleaning supplies

There are a variety of other supplies that might come in handy when repairing and maintaining a VCR.

☐ Brushes let you dust out dirt. Any good-quality painter's or artist's brush will do. Stock a small and a wide brush so you can tackle all jobs.

☐ Contact cleaner enables you to clean the electrical contacts in the deck. The cleaner comes in a spray can, but you can apply it by spraying the cleaner onto a brush and whisking the brush against the contacts.

☐ Cotton swabs help you soak up excess oil, lubricant, and cleaner. Swabs are available in quantity at any drugstore.

☐ Sponge-tipped swabs are like cotton swabs but leave no lint behind and are ideal for use when cleaning rotary heads. The swabs can be purchased at Radio Shack, most electronics stores, and at the cosmetic counter at the drugstore.

☐ Small chunks of untreated, virgin chamois or deerskin are a suitable alternative to sponge-tipped swabs. Purchase the chamois at an auto parts store and cut a portion of it into small one-inch squares.

☐ Orange sticks (from a manicure set) and nail files let you scrape undesirables off circuit boards and electrical contacts.

☐ The eraser on a pencil goes a long way to rub electrical contacts clean, especially ones that have been contaminated by the acid from a leaking battery.

☐ Modeling putty (for plastic models) can be used to mend cracks and chips on the plastic exterior of the VCR cabinet.

☐ Contact cement, white glue, and other common adhesives are good for repairing broken plastic and metal parts.

□ A small magnet makes it easier to pick up screws and ferrous metallic objects that have been accidentally dropped into the deck.

□ A fluid called Regrip helps rejuvenate rubber parts, used in idler wheels and belts. Alcohol should not be used to clean rubber parts, as it can dry them out. Don't use any product that leaves a tacky residue (such as Non-Slip, available at Radio Shack).

See appendix A for a list of sources for these products.

## Capstan roller rejuvenator lathe

Common problems of about any VCR is capstan pinchroller and idler tire wear. The capstan pinchroller is used to provide pressure against the capstan shaft. The shaft and roller pull the tape through the transport mechanism during play and record. If the rubber in the roller becomes old and brittle, playback and recording is impaired.

The idler tire is used to propel the take-up and feed spindles during playback, rewind, and fast forward. If the tire becomes old, brittle, or dirty, it can slip, and performance of the deck is impaired. With many VCRs, an old idler tire can cause tape spillage, which can ruin the tape and cause jams. Spillage occurs because the take-up reel isn't moving fast enough to wind up the tape that's being fed through the machine during playback.

You can replace the capstan roller and idler tire, but a new part could be outrageously expensive (some VCR manufacturers and retailers might charge as much as $5 to $10 for the parts), or it could be hard to locate. In most cases, you can rejuvenate the rubber of the tire by carefully scraping the outside with a fingernail emery board, as discussed more fully in chapter 5. You can redress the capstan roller the same way, but a better approach is to use a homemade "lathe" constructed out of a drill motor and spare hardware parts.

Figure 4-10 shows how to construct the makeshift lathe and how to attach the capstan pinchroller. Use the proper diameter of bolt to accommodate the hole in the roller. Almost any variable-speed drill motor will work (use the drill at a relatively slow speed). A motorized portable screwdriver also can be used, but its chuck might not accept many different bolt sizes. Chapter 5 details the proper use of the lathe.

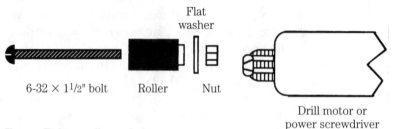

Flat
washer

6-32 × 1¹/₂" bolt    Roller    Nut

Drill motor or
power screwdriver

To use: Tighten roller on bolt
with nut; attach stem of bolt in
drill motor chuck.

■ **4-10** *How to use a drill motor to construct a pinchroller rejuvenator lathe. Use a larger or smaller bolt and nut to accommodate the shaft size of the roller or tire you are cleaning.*

## Using test tapes

Test tapes enable you to put your VCR through a series of diagnostic procedures that examine the deck's ability to play back sound and picture successfully. The tape tests the machine's alignment (the picture will jitter if components are out of alignment) and playback functions. Test tapes are available commercially but can be expensive for the part-time VCR maintenance enthusiast. Most VCR manufacturers (names and addresses are supplied in appendix A, "Sources") provide test tapes. Write them for a current catalog and price list.

Unless you are repairing VCRs for other people, you don't really need a commercial test tape. You do not need to realign the internal components of the VCR to meet some industry standard; as long as your VCR plays your old tapes with no problems and can accept prerecorded tapes, the alignment is perfectly acceptable.

Any tape you made on your deck prior to the occurrence of any problems can be used for testing. It's a good idea, actually, to make a test tape when the deck is brand new. Store the tape in a cool, dry place in case you need it at some later date. Prerecorded movies are another good candidate for test tapes.

The better prerecorded tapes (of major movies, not off-the-wall art films) are typically made on high-quality tape duplicators that are constantly maintained in top performance. If your VCR successfully plays a tape recorded on one of these machines, you can be fairly assured that its alignment and playback circuits are within acceptable design limits.

Take note, however, that if your deck is "eating" or otherwise damaging tapes, you should not risk the expense of using a prerecorded movie. The tape might get severely damaged by the deck, and if you are renting the movie, you will need to reimburse the store for the ruined merchandise. Use prerecorded movie tapes only when you are certain that the deck won't cause physical damage.

# Cleaning and preventive maintenance

THE SAYING, "AN OUNCE OF PREVENTION IS WORTH A POUND of cure," applies in many walks of life, and it definitely goes for videocassette recorders. A few well-spent moments keeping your VCR in top-notch condition goes a long way in keeping it healthy. You'll save money and enjoy your video investment more. A clean VCR is a happy VCR.

In this chapter, you'll learn how to give your VCR a routine preventive maintenance checkup. This check includes the deck itself, the remote control, and even your collection of tapes. Also, you'll learn how to test the machine for proper operation and how to repair moderately damaged tapes.

## Frequency of checkup

The PM (preventive maintenance) interval for your VCR varies, depending on many factors. Under normal use, you'll want to give the deck a PM checkup every 6 to 12 months. *Normal use* means a VCR operated in an average household environment—not on a factory floor or at the beach.

You might need to perform the PM checkup on a more timely basis if your deck is subjected to environmental extremes or is exposed to heavy doses of dust, dirt, water spray (especially salt water), airborne oil, and sand. A chart of suggested PM intervals for machines that receive light, medium, and heavy use is shown in Fig. 5-1 on the next page.

How do you know if your VCR needs a preventive maintenance checkup? Good question. Experience is your only guide, but if you have not yet gained the experience in how often the deck needs routine maintenance, you can't judge if the service interval is required. As you become acquainted with your videocassette recorder, you

**Light Use**

| PM (Months) | 3 | 6 | 9 | 12 | 15 | 18 | 21 | 24 | 27 | 30 | 33 | 36 |
|---|---|---|---|---|---|---|---|---|---|---|---|---|
| Clean/Dust Exterior/Interior | □ | □ | ■ | □ | □ | ■ | □ | □ | ■ | □ | □ | ■ |
| Clean Heads | □ | ■ | □ | ■ | □ | ■ | □ | ■ | □ | ■ | □ | ■ |
| Lubricate Transport | □ | □ | □ | □ | □ | ■ | □ | □ | □ | □ | □ | ■ |
| Clean Rubber Parts | □ | □ | □ | □ | □ | ■ | □ | □ | □ | □ | □ | ■ |
| Clean Remote Battery Terminals | □ | □ | □ | ■ | □ | □ | □ | ■ | □ | □ | □ | ■ |
| Clean Connectors | □ | □ | ■ | □ | □ | ■ | □ | □ | ■ | □ | □ | ■ |

**Medium Use**

| PM (Months) | 3 | 6 | 9 | 12 | 15 | 18 | 21 | 24 | 27 | 30 | 33 | 36 |
|---|---|---|---|---|---|---|---|---|---|---|---|---|
| Clean/Dust Exterior/Interior | □ | ■ | □ | ■ | □ | □ | □ | ■ | □ | ■ | □ | ■ |
| Clean Heads | □ | ■ | □ | ■ | □ | □ | □ | ■ | □ | ■ | □ | ■ |
| Lubricate Transport | □ | □ | □ | ■ | □ | □ | □ | □ | □ | □ | □ | ■ |
| Clean Rubber Parts | □ | □ | □ | ■ | □ | □ | □ | □ | □ | □ | □ | ■ |
| Clean Remote Battery Terminals | □ | □ | ■ | □ | □ | ■ | □ | □ | ■ | □ | □ | ■ |
| Clean Connectors | □ | ■ | □ | ■ | □ | ■ | □ | ■ | □ | ■ | □ | ■ |

**Heavy Use**

| PM (Months) | 3 | 6 | 9 | 12 | 15 | 18 | 21 | 24 | 27 | 30 | 33 | 36 |
|---|---|---|---|---|---|---|---|---|---|---|---|---|
| Clean/Dust Exterior/Interior | □ | ■ | □ | ■ | □ | ■ | □ | ■ | □ | ■ | □ | ■ |
| Clean Heads | ■ | ■ | ■ | ■ | ■ | ■ | ■ | ■ | ■ | ■ | ■ | ■ |
| Lubricate Transport | □ | ■ | □ | ■ | □ | ■ | □ | ■ | □ | ■ | □ | ■ |
| Clean Rubber Parts | □ | ■ | □ | ■ | □ | ■ | □ | ■ | □ | ■ | □ | ■ |
| Clean Remote Battery Terminals | □ | □ | ■ | □ | □ | □ | □ | □ | ■ | □ | □ | ■ |
| Clean Connectors | ■ | ■ | ■ | ■ | ■ | ■ | ■ | ■ | ■ | ■ | ■ | ■ |

**Your Machine**

| PM (Months) | 3 | 6 | 9 | 12 | 15 | 18 | 21 | 24 | 27 | 30 | 33 | 36 |
|---|---|---|---|---|---|---|---|---|---|---|---|---|
| Clean/Dust Exterior/Interior | □ | □ | □ | □ | □ | □ | □ | □ | □ | □ | □ | □ |
| Clean Heads | □ | □ | □ | □ | □ | □ | □ | □ | □ | □ | □ | □ |
| Lubricate Transport | □ | □ | □ | □ | □ | □ | □ | □ | □ | □ | □ | □ |
| Clean Rubber Parts | □ | □ | □ | □ | □ | □ | □ | □ | □ | □ | □ | □ |
| Clean Remote Battery Terminals | □ | □ | □ | □ | □ | □ | □ | □ | □ | □ | □ | □ |
| Clean Connectors | □ | □ | □ | □ | □ | □ | □ | □ | □ | □ | □ | □ |

■ **5-1** *Preventive maintenance schedule for VCRs subjected to light, medium, and heavy use.*

will get to know when it needs a checkup. In the meantime, here are some clues that can point you in the right direction:

☐ The outside of the deck is covered with caked-on dirt, dust, or other grime. If the outside is dirty, so is the inside.

□ The summer has come and gone and you feel your camcorder has seen all the good times with you. Was your camcorder buried in the sand at the all-night beach party? Even if it still plays, it's no guarantee that it's having an easy job at it.
□ The picture and sound just aren't what they used to be. Something is amiss inside, but not seriously enough to entirely impair playback.
□ The player makes unusual sounds during tape loading and playback.

## A word of caution

It's important to remember that unless you are an authorized repair technician for the brand of deck you own, taking it apart to perform the preventive maintenance procedures outlined in this chapter will probably void the warranty. Most VCRs have a warranty period of one year or so (some more, some less), and you might as well take advantage of it. There is usually no reason to perform a PM checkup within the warranty period.

One exception to this is if the machine has been accidentally damaged by water, dirt, or sand, and although it still works, you'd feel better if it were cleaned. Factory warranty service does not cover this type of cleaning, and you'd pay handsomely for it. The cleaning and "repair" cost could well exceed the price you paid for the machine. In this case, there's nothing stopping you from doing the work yourself. If your VCR is subjected to heavy abuse, check the first-aid procedures in chapter 6.

## Personal safety

Once again, be aware of the potential dangers that lurk inside your VCR. If yours is an ac-operated home model, removing the cover exposes 110 V (or more) of alternating current. Given the right circumstances, the current involved can kill you. Remember to unplug the unit at all times when you are not actually testing its operation. Unless otherwise specified, all maintenance and repair procedures are made with the VCR off and unplugged from the wall socket. Unplugging the unit is an important step. Even with the power switch off, electricity is still present at the terminals near the unit's power supply unless the VCR is unplugged.

Warning signs that appear on the back of ac-operated VCRs (and in their manuals) say there are potentially dangerous voltages inside. When you see the triangle and lightning bolt symbol, as shown in Fig. 5-2, use caution. The other warning sign shown in the figure, the triangle and exclamation point, says that you should refer to the owner's manual before operating the equipment.

**Warning: Do not perform the steps in this chapter unless you feel confident in your ability and have observed all safety precautions. Take your time and think about every step. If you don't feel you have time for the general cleaning and maintenance procedures, put it off until another day when you have more opportunity. Rushing things will surely lead to disaster.**

■ **5-2** *Warning signs on the back or bottom of the VCR.*

## General maintenance

Before you go poking around inside the VCR, first take a close look at the outside. You can perform these preventive maintenance steps on a more regular basis because they do not require removing the cover of the player.

### Operation manuals

An important measure in your efforts to minimize VCR troubles is to read the instruction manuals that came with your equipment. This step might sound a bit obvious, but you'd be surprised how many video enthusiasts never take the time to read the operating manuals thoroughly. The manufacturers of the equipment include the manuals so you can get the most out of your expensive purchase. Keep them at hand, and refer to them whenever you have a question.

### Exterior dust and dirt

One of the first items on your system maintenance list should be routine external cleaning. Every few days, take a dry cloth and wipe each component of your video system. Don't use dusting sprays; these actually attract dust, luring dirt back onto your equipment. Use a soft, sable painter's brush for those hard-to-reach places. Be sure to clean the ventilation slots as these are favorite hiding places for dust.

If you need to get rid of stubborn grime, apply a light spray of regular household cleaner onto a clean rag; then wipe with the rag. Never apply the spray directly onto the deck because the excess can run inside. Never apply a petroleum- or acetone-based solvent cleaner as it might remove paint or melt the exterior plastic parts. Some plastics, when in contact with solvents, give off highly toxic fumes.

## Cables

The cables and connectors in your video system can make your VCR sneeze and wheeze, so make sure they're on tight and that none of them have become damaged. Many owners of VCRs have a tangle of cables behind their TV, and just one bad wire or connector in the bunch can cause grief.

Most important is coaxial cable maintenance. Coax cable consists of a solid center conductor surrounded by a heavy plastic called the *dielectric*. Around this is a mesh of braid that not only acts as a second conductor but as a shield against external interference. The entire cable is insulated with plastic insulation on the outside.

137

Coax cannot cope well with torture, yet it's often subjected to twisting, tugging, kinking, and tight looping. If you've got excess, wind it loosely; never bundle it up like extra speaker wire. Be certain that none of your coax is in foot traffic. That includes routing it under carpets. Walking over coaxial cable can deform the dielectric, creating ghosts and loss of signal. Tripping over the cable can tug on the connectors, or worse yet, pull your equipment to the floor.

While inspecting the cabling, make sure to remove any kinks that might cause signal loss and ghosting. Look closely at the connectors attached to the ends of the cable to make sure they are on tight and that the center conductor has not been broken or bent (see Fig. 5-3 on the next page).

If you suspect that a cable might be bad or causing interference and hum, try a replacement to see if the problem is solved. Alternatively, you can use a volt-ohmmeter, as shown in Fig. 5-4 on the next page, to test the continuity of the cable. Attach the test leads to the center conductor of both ends and take a reading. The meter should read 0 $\Omega$. Do the same for the outer conductor.

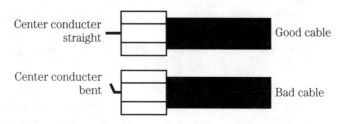

Center conductor straight ............................................... Good cable

Center conducter bent ..................................................... Bad cable

■ **5-3** *The center conductor on a good coaxial cable should be straight and shiny.*

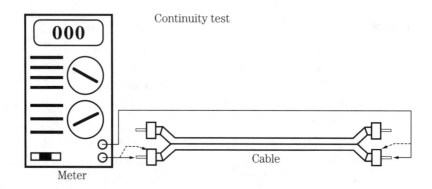

Continuity test

000

Meter

Cable

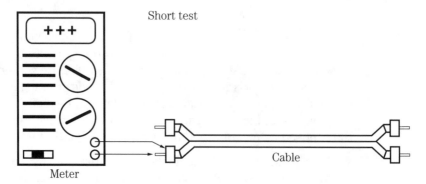

Short test

+++

Meter

Cable

■ **5-4** *Use a meter to test the condition of cables in your video system. You can test for continuity and shorts by dialing the meter to the ohms (or resistance) setting.*

Also check that the cable has not shorted by applying the test leads to the inner and outer conductors on one end of the cable (be sure not to hold onto the metal probes of the test leads or you might get false readings). The measurement should read infinite ohms (open circuit).

# Internal preventive maintenance

To get inside the deck, remove the deck from its location in your video system. Disassemble the VCR only in an open, well-illuminated work area, as described in more detail in chapter 4.

## Disassembly

Disassembly of most VCRs is simple and straightforward. The typical ac-operated home unit is composed of a top and bottom piece; you loosen two or more screws and remove the top portion of the cabinet. The screws are usually located at the rear and sometimes the side of the deck, as shown in Fig. 5-5. The bottom panel, which does not need to be removed unless you are servicing the capstan drive and threading mechanisms, is detached by removing the screws on the bottom (see Fig. 5-6). Note that in some early Sony and Zenith Beta decks, you must first remove the tracking control knob before removing the front cover.

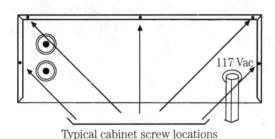

Typical cabinet screw locations

■ **5-5** *Common locations for screws on the back of the VCR.*

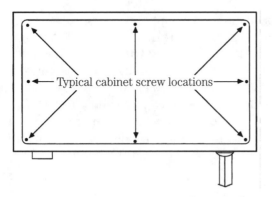

Typical cabinet screw locations

■ **5-6** *Common locations for screws on the bottom of the VCR.*

Do not loosen every screw you see, especially those on the bottom center portion of the cabinet, because they might be anchoring some internal components. Loosen one screw at a time and attempt to remove the cover (or least see if it is coming loose). Run your fingers along the edges of the cabinet to see where it is still catching. If a screw seems to hold an internal component in place, retighten it and try another.

The cabinet screws on some VCRs are marked with an arrow or paint to indicate that they should be removed. If the screws on your VCR are so identified, remove only the ones that are marked. Leave the others. Top-loading VCRs require that you remove the top plate before removing the cabinet cover. The top plate is held in place by two or four screws.

Save the screws in a cup or container (or your fishing tackle box). If the screws are different sizes (they probably will be), note where they came from for easier reassembly. Keep washers, grommets, spacers, and other hardware with their related screws. The additional hardware has been used for a reason and should not be omitted when the deck is reassembled.

If necessary, write a list of the parts you remove and the order in which they were removed. If a certain screw has a rubber washer, a metal washer, and a locking washer on it, write down the order in which they are assembled on the screw. Carefully inspect the parts you remove to determine if they are aligned or oriented in a certain way. Indicate the alignment or orientation on your sheet so you are sure to duplicate it when you put the machine back together again.

With the screws removed, gently lift off the top cover, being careful not to disturb the internal components or wiring. You might need to slide the cover back before taking it off, as shown in Fig. 5-7, if the cover tucks under a lip in the front panel.

Once the cover is off, do not attempt to disassemble anything else inside the machine except where noted in the following paragraphs. This precaution is especially true of front-loading VCRs. The cassette-lift mechanism can be difficult to reassemble if you don't have the service manual to refer to. In addition, loosening the wrong screws can skew the orientation and position of critical components. The altered orientation might require a service manual (which is expensive) and a special alignment jig (which you cannot easily get for yourself) to correct the mistake.

## Static electricity

Static electricity in your body can harm the integrated circuits and other components on the main circuit board. Therefore, do not touch any part unless you have first drained the static electricity from your body. If you plan to handle the circuit boards, use a grounding strap around your wrist. The strap is available at most

■ **5-7** *Remove the top cover of the VCR by lifting from the back and sliding the top out from under the lip in the front panel.*

electronics supply stores. Follow the instructions supplied with it for proper use.

## Preliminary inspection

Take a few moments to acquaint yourself with the inside of the deck. Note the position of the video head drum, the loading and threading mechanisms, the main circuit board, the auxiliary boards for the front panel, the indicator, the tuner, the video section, the routing of the cables, and the incoming power supply lines. Avoid touching anything, especially the video head drum, unless you absolutely must. Figures 5-8 through 5-13 on the following pages show the inside of a typical VCR. Figure 5-14 on page 145 shows the view of the bottom of the VCR, the capstan drive mechanics, and the threading gears and associated mechanism.

If you like, draw a picture of the inside of your VCR, and note where things belong and how they work. When you are well acquainted with your VCR, you can service it better and more confidently. Note especially the location and appearance of all the belts, tires, pulleys, fuses, and other mechanical parts. You might want to slip a tape into the deck and watch it load and thread. Note the action of all the parts.

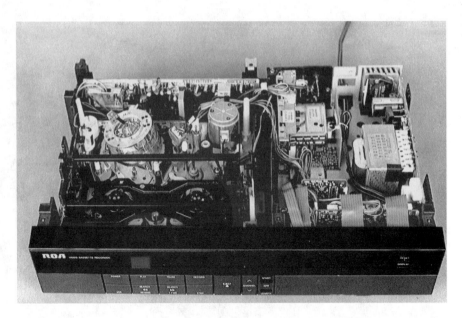

■ **5-8** *Top view of the VCR with the cover removed.*

■ **5-9** *The video-head drum.*

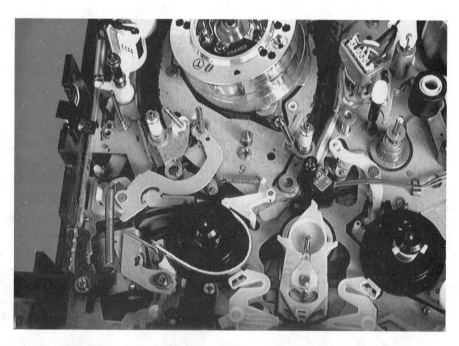

■ **5-10** *The transport mechanism.*

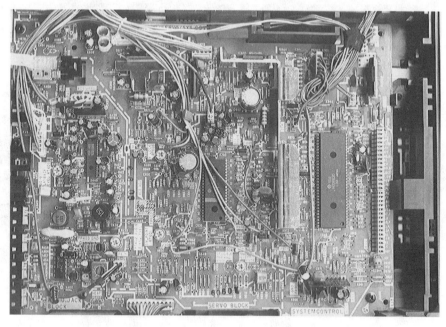

■ **5-11** *The main printed circuit board (underside of deck on this model).*

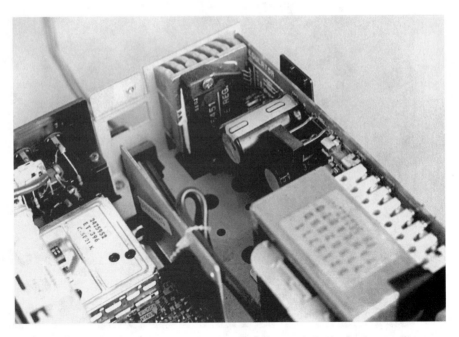

■ **5-12** *The power supply section (the transformer is in the foreground).*

■ **5-13** *The RF section. The enclosed module is the RF modulator.*

**■ 5-14** *Bottom view of the transport mechanism showing the capstan, flywheel, and belts.*

One word of caution here: avoid operating the VCR in any position but upright. Don't turn the VCR over or on its side, especially during threading. The tape can spill out and become jammed in the works. Operating the VCR in an abnormal position also poses greater stresses on the motors and mechanical parts, which could lead to premature failure.

While the top (or bottom) of the deck is off, be particularly wary of the power supply board or power supply section on the main PCB (printed circuit board). The filtering capacitors (which usually look like tall cans) retain some charge even when the player is turned off and unplugged. Avoid touching the leads of the capacitors and make sure that you do not short the leads together. You can damage the capacitor and other components on the board.

Front-loading decks have more complex loading mechanisms than their top-loading cousins. There is more to look at in a front-load machine and consequently more points that might need preventive care.

Keep in mind there is little, if any, adjustment or calibration that you can—or should—do once inside the deck. Potentiometers on the circuit boards are factory set and require an oscilloscope or special test circuitry for proper adjustment. The nature of VCRs inherently leaves few internal controls you can safely tweak.

In addition, the latest models are designed with few alignment points. Older models—those manufactured before 1981 or 1982—have service points for fine tuning the take-up and supply-reel torque, pinchroller pressure, back tension, etc. Today, VCRs are designed with engineering tolerances built-in—these adjustments are rarely found and no longer need to be made.

## Tires, pinchrollers, and belts

Visually check the tires and belts used in the reel spindles, the tape loading mechanisms (gears are sometimes used here), idler tires, and capstan pinchroller. If worn or cracked, they should be cleaned or replaced.

To examine the rubber piece, you might need to stretch or twist it, as you would an automotive fan belt. Look for a glazed appearance on the rubber; that indicates wear. Cracking is a sign that the rubber is brittle and dry and is no longer doing its job. It must be replaced. An oily film on the rubber can be an indication that the VCR was excessively lubricated, and that oil or grease from the mechanical sections of the deck have migrated to other components. The oil must be removed to restore proper traction. Healthy rubber has a suppleness to it. With experience, you will learn to tell the difference between strong, healthy rubber parts, and weak, sickly ones.

Idler tires and pinchrollers can develop flat spots, especially if the VCR has not been used for some time. You can visually inspect for flattening; if it is only minor, rejuvenating the rubber—as discussed in this chapter—can repair the damage. If the tire or roller is excessively flat, it must be replaced.

You can replace belts yourself. Get the service literature for your player from the manufacturer, identify the part by its part number, and order it from the manufacturer. Belts and idler tires (you replace the tire, not the complete mechanism) can often be ordered simply by measuring the tire diameter and size. Dentist's picks, which you can buy at most electronics and surplus outlets, let you more easily remove and install belts that are hard to reach. Figure 5-15 shows a typical belt configuration as used in the capstan drive mechanism. Note that the capstan motor drives several sections of the deck.

Rubber rollers and belts that are slipping can be cleaned with a special nonslip cleaner designed to rejuvenate rubber. The nonslip cleaner is available at many electronic supply stores, and can

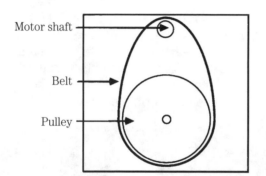

Motor shaft

Belt

Pulley

■ **5-15** *The typical arrangement of the belt between the capstan motor and the pulley that drives the capstan flywheel and capstan shaft.*

sometimes be purchased through a typewriter repair shop (they use a similar mixture to clean the rubber platen of the typewriter). Apply it to the belt or roller with a cotton swab or sponge applicator, as shown in Fig. 5-16. Wait a few minutes, then vigorously wipe it off with a cotton swab. Use a disposable fingernail file (emery board) to "sand" the old, top layer. Scrape the flat of the board across the rubber in even, gentle strokes. Don't do more than one stroke at any one spot, or you might make the tire or roller flat.

You can achieve better cleaning results of pinchrollers (and similar parts) by using the homemade motorized "lathe" discussed in

147

■ **5-16** *Clean the idler tire with a swab soaked in cleaner or a rubber rejuvenator.*

chapter 4. The lathe in operation is shown in Fig. 5-17. To use the lathe, mount the pinchroller on the bolt. By keeping several sizes of bolts and nuts handy, you can accommodate the shaft sizes of a variety of tires and capstans. Tighten the nuts so the roller does not spin on the shaft of the bolt. Do not overtighten because you might break the pinchroller or deform the rubber.

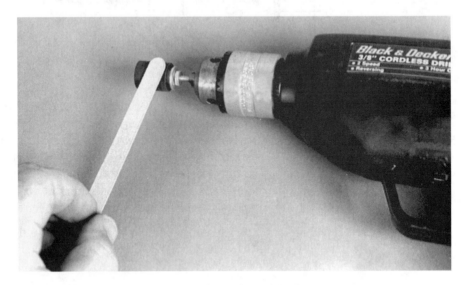

■ **5-17** *A lathe restoring the surface of a pinchroller.*

Rest the drill motor or motorized screwdriver on a flat, hard surface, and turn it on. Reduce the speed to a moderate 100 to 300 *rpm* (revolutions per minute). Lightly rub an emery board against the rubber of the roller while it is spinning. Do not press hard; make sure that you run the board evenly across the surface of the rubber, as shown in Fig. 5-18. Periodically stop the lathe and inspect the part. Although the exact diameter of the roller is not critical, you should not remove more than a very thin layer of rubber.

Avoid the use of solvent-based cleaners on rubber parts, as these chemicals can prematurely dry up the rubber, or worse, melt it. Never apply a lubricant of any type to a tire, roller, or belt because that will defeat its purpose. If the rubber squeaks, it's an indication that something is not properly aligned or that the rubber needs cleaning or replacement.

Tires, belts, and rollers press against plastic and metal parts. After years of use, rubber can cake up on these parts, causing a loss of traction. While you are cleaning the idler tires, belts, and pinchroller, clean the surfaces these parts come into contact with. Use alcohol or a nonpetroleum-based solvent to clean the part. Caked-on rubber

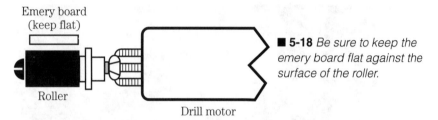

■ **5-18** *Be sure to keep the emery board flat against the surface of the roller.*

that refuses to come off can be removed with an orange stick borrowed from a manicure set. Pay particular attention to:

☐ Capstan shaft
☐ Reel spindle rims
☐ Idler tire driver
☐ Capstan motor pulley (underside of deck)
☐ Capstan shaft flywheel (underside of deck)
☐ Threading/loading pulleys

## Brakes and torque limiters

The supply and take-up reels of VCRs are fitted with brakes that prevent free movement during normal recording and playback. Some VCRs use a number of different brakes: a band-shaped brake is used on the supply reel to compensate for tape slack; a brake on each reel stops the spindles from turning after fast-forwarding or rewinding; brake pads on both spindles limit the torque of the reels, preventing undo tape tension. A closeup of the brake system in a VHS deck is shown in Fig. 5-19 on the next page.

Inspect the condition of the brake pads. The band brake uses a strip of felt—be sure the felt hasn't worn off. The other brakes use rubber as the brake pads. Inspect the rubber and use Regrip to rejuvenate it, if necessary. Clean the contact surfaces of the reel spindles with alcohol to get rid of any caked-on deposits.

## Topical cleaning

Dust, dirt, and nicotine from cigarette smoke can accumulate in the interior of the VCR and should be removed. Use a soft brush to wipe away excess dust and dirt. If there is a lot of sediment, use a small hobby vacuum cleaner.

Remaining dust and other junk can be blown out of the inside of the deck. Purchase a small can of compressed air at a photographic shop (about $3) and liberally squirt it inside the machine. That should free about everything that shouldn't be there. Remember to keep the can upright to prevent the propellant from

**5-19** *The brake on the supply reel spindle.*

dripping out. As much as possible, position the air flow so debris is blown out of the VCR, not deeper inside!

Avoid using household cleaning sprays because these not only can leave a residue that impairs the proper operation of the deck, but they are water-based and might short out the circuitry. Do not use a cleaner that contains a solvent of any type, and do not clean the machine with an oilless lubricant such as WD-40. This spray is nonconductive and will cause the deck to fail completely.

You can clean nicotine sediment and caked-on dirt using a cleaner recommended for application on mechanical and electronic components. Figure 5-20 shows such a cleaner being used on the main PCB (apply the cleaner only when the deck has cooled down after use). Some cleaners have a built-in degreaser that disperses oil buildup; these are fine for use in your VCR, as long as they are designed for direct application onto electronic parts. Whatever cleaner you use, test it on another similar surface to make sure it leaves no residue after drying (which usually takes less than 15 seconds). If the cleaner leaves a noticeable residue, it is unsuitable for use inside your deck.

The proper cleaner is available at most electronics stores and usually comes in spray form. The spray comes out fast, so you can use

**■ 5-20** *Spraying the main PCB with a nonresidue cleaner/degreaser.*

the pressure to dislodge stubborn grime. Most cans come with a spray-nozzle extension tube. Attach the tube to the spray button and squirt into hard-to-reach places.

Remember that prior to cleaning, be absolutely sure the VCR is unplugged from the wall outlet. Never use the spray where there is live current. You could receive a shock or start a fire. After cleaning, wait at least three minutes for all the cleaner to evaporate before applying power to the deck.

## Electrical contact cleaning

It's rare that the electrical contacts inside the deck need cleaning. However, exposure to outside elements can quickly oxidize plated electrical contacts (now you know why you should never place your VCR near an open window where it might be exposed to the damp outside air). The contacts also can be contaminated by an accumulation of cigarette smoke.

If you see a heavy amount of oxidation or nicotine, carefully disconnect the wire leading to the contact and clean the contact with a commercially available contact cleaner. Clean the exposed part of the wire in a similar fashion. If there is a heavy buildup of oxidation, scrape it off with an orange stick or nail file and then use the cleaner.

Most contact cleaner is in spray form, and the spray might squirt to an overly large area. If you don't want overspray, spray the cleaner into a cotton ball or swab, then use the cotton as an applicator. Be sure not to leave lint from the cotton on the contacts. If cotton threads are a problem, purchase some sponge-tipped applicators. Most electronics and many record stores sell them. For very hard-to-reach places, use the extension nozzle tube.

Some circuit boards in the VCR are connected to the main board by an edge-card connector. If you remove the circuit board for servicing or cleaning, be sure that you don't touch the plated *lands* on the board or the connector pins on the main PCB. Not only is there a chance of damaging a component with a shock of static electricity from your body, the oils in your skin will act as a corrosive acid, diminishing the effectiveness of the electrical contact.

## Video heads

VCRs use spinning magnetic heads to record and play back video information to and from the tape (VHS hi-fi decks also use rotating audio heads, and a few VCR models have rotating flying erase heads). The assembly and the way the tape wraps around the video head drum is shown in Fig. 5-21. Like any other magnetic head, the video heads will eventually become dirty (or clogged) due to an accumulation of oxide coating shed from the tapes. The heads also can become dirty by direct contact with dirt, ash, grime, or other contaminates on the tape surface.

Dirty heads also can create excessive dropout. Dropout looks like salt-and-pepper flecks on the screen. Depending on the amount of dropout, the flecks can be anything from a light snow to a heavy blizzard. Routine and regular cleaning of video heads will ensure carefree operation and optimum performance.

The best method of cleaning video heads is a debatable question and is one that only you can decide for yourself. Covered here are the two most accepted methods; you can decide which one suits you the most. You might even want to alternate between the two methods, as each one has its own distinct advantages.

### Cleaning cassettes

By far, the easiest and fastest way to clean video heads is by using a commercially manufactured cleaning cassette like the one pictured in Fig. 5-22. Use is simple: just pop the cassette into the machine, play it for the recommended time (usually less than 20 seconds),

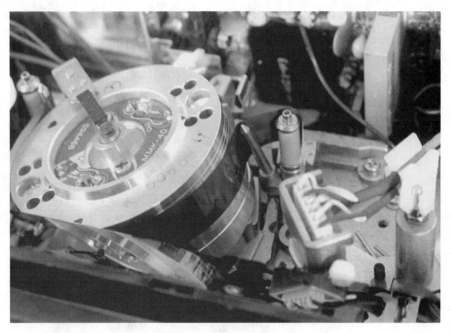

■ **5-21** *The tape wrapped loosely around the video drum shows the path of the tape through the mechanism of the VCR.*

■ **5-22** *A commercially available wet/dry head-cleaning system.*

and put it away until the next time. A few of the cleaning cassettes display a message on the screen when cleaning is complete.

Some head-cleaning cassettes (particularly the dry type) work by abrasive action, so naturally you'll want to take it easy and avoid using these tapes until you really need them. It is possible to wear down a video head by using a cleaning cassette, so follow directions carefully and write down the date whenever you use the tape. If you notice that you're cleaning the heads more than every one month or so, you might have other problems, like an old tape that's shedding too much of its oxide coating. There is more information about this in the chapter.

Other cleaning cassettes use the *wet* or *wet/dry* method. Before using the cassette, apply a drop of a liquid cleaner to the tape. When you insert the cassette, the wet tape cleans the heads and any other metal or plastic parts it touches. The problems of overuse and abrasion are not as great with this type of cleaning tape, but use it carefully nonetheless.

Normally, head-cleaning cassettes do a fair job at cleaning the spinning video heads as well as the stationary heads and other metal parts inside the VCR. A few brands do not clean all the stationary heads, however, nor the entire tape path.

True, the stationary heads do not need as stringent a cleaning as the video heads, but an excessive buildup of dirt and oxide can lead to sound distortion, inadequate erasing, and faulty picture synchronization. If your favorite cassette doesn't clean the entire innards (as described in its instruction manual), consider supplementing it with another brand or method.

## Manual cleaning

The other video-head cleaning method involves manually cleaning the heads. You need two supplies: cleaner and cleaning pads. A Freon TF equivalent and alcohol mixture (available at any electronics store) or other recommended video-head cleaning solution is ideal for cleaning video heads. Isopropyl alcohol also can be used, as long as it is technical grade or medical grade—97 percent alcohol or more. The higher the percentage of alcohol, the less water is in the mix. The water takes longer to evaporate, making maintenance more difficult, and it can prevent adequate removal of oxide and other contaminants. Don't use a cleaning solution designed for audio heads. These sometimes have lubricants that aren't needed or desired for video heads.

Be especially wary of head cleaners that you spray on, though these can be used. The cleaning agent in these is a fluorocarbon like Freon, or a solvent with similar cleaning properties of Freon, which is very cold (it actually reaches its boiling point when it hits the air). The sudden blast of cold can easily crack a warm or hot video head, rendering it useless. If you use a spray-on head cleaner, be sure the VCR has cooled completely before you spray.

Apply the cleaner with a chamois-tipped swab or pad only—never use cotton swabs on the video head because the lint can come off and become entangled in the head mechanism. You also can use a piece of clean, virgin chamois cut into a convenient square-inch piece. When cleaning, thoroughly wet the swab and gently rub it sideways on the head, as shown in Fig. 5-23. Never use an up-and-down motion—you will surely break the delicate heads right off. Most VCRs have several video heads, so be sure you get them all.

■ **5-23** *Cleaning the video heads with a chamois-tipped applicator. Remember to fully wet the chamois with the cleaner.*

One way to clean the heads safely is to apply the swab (or sponge or chamois) to the side of the head drum and rotate the top portion of the drum manually with your fingers (see Fig. 5-24 on the next page). Apply only moderate pressure on the cleaning pad. When using a chamois pad, you will feel the head pass under your finger.

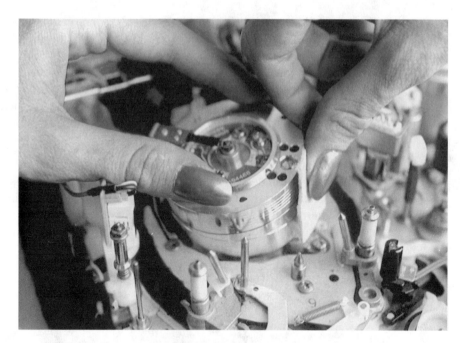

■ **5-24** *Press a small square of chamois against the head drum. Move the heads back and forth by rotating the top part of the drum with your other hand.*

After cleaning, wait at least five minutes before inserting a tape and playing it. The waiting period is important; it allows the cleaner to evaporate fully. If you play a tape and the heads are still wet, they will get dirty again, and the oxide residue will be even harder to remove.

Throw the cleaning swab or chamois away after use. Although it might still seem clean, reusing it can impair the effectiveness of future cleanings.

Sometimes, the heads can be so dirty that ordinary cleaning won't remove the oxide buildup. If the picture won't clear after two or three cleanings (wait the recommended time between each cleaning), try a dry-type cleaning tape. The lapping action of the tape will likely scrub off the oxide and clear up the video.

Note that all VCRs (except for some formats of camcorders, like BetaMovie) use at least two heads. Depending on the model, format, and whether it has hi-fi stereo, there can be eight (and sometimes even more!) heads on the drum. You can sometimes determine the function of each head by reading the labeling on the top of the head drum; otherwise, consult the service manual for your machine. In any case, be sure that you get all the heads. The entire process might take 5 or 10 minutes if your deck is equipped with four or more heads.

## What to do with hard-to-clean heads

Sometimes video heads are extremely hard to clean. No matter how much you scrub with a cleaner, or use a cleaning cassette, the heads still remain a little dirty. This situation occurs most often when the dirt has built up over a long time, and the grime has been polished in to the smooth surface of the head.

If the video heads in your VCR prove resilient against cleaning, your best defenses are:

☐ Clean the heads manually, with an approved video head cleaner and lint-free swabs, as detailed previously. Avoid using cleaning cassettes.
☐ Wait at least five minutes between cleanings, and before you test the job with a videotape. Waiting is important: it allows the water and solvent of the cleaner to evaporate completely. As mentioned earlier, playing a videotape too soon after cleaning can cause even worse build-up, as new grime literally becomes cemented to the surface of the video heads.
☐ Consider the tapes you use to be the cause of the dirt. If the heads stay clean for only a little while, they might be in old or poor-quality videotapes. Use different tapes (preferably name brand) to see if the problem goes away.

## Cleaning other heads

While you're inside the machine cleaning the video heads, get all the other heads as well. Include the full-erase head and the main head unit, which includes the control-track, audio, and (possibly) the audio-erase heads. Be sure to clean thoroughly each metal part that comes into contact with the tape including guide rollers, impedance rollers, treading guides, and even the entire video-head drum surface (this removes gunk that can cause friction on the tape). You can use the cleaner on hard plastic parts (oxide buildup is especially noticeable on these), but avoid applying the cleaner on soft rubber parts. These require a special rubber cleaner (available at most electronics stores).

## How often to clean heads

Some experts say you should clean the heads only when the play-back picture shows a definite loss in clarity or shows an increase in snow. Others say to clean every 75 to 100 hours of use. Still others claim you should do it every 25 to 35 hours of use. Who's right? They all are. It's what you're comfortable with, what you need, and what the manufacturer recommends for its product.

The thing to remember is that the heads usually do not require cleaning every day or even every week. With a VCR that receives moderate use, cleaning once every month or two is reasonable.

### Inspecting the auto head cleaner

A growing number of VCRs come with an automatic head cleaning feature. This feature consists of a head cleaning band or arm that makes momentary contact with the spinning video heads each time a tape is removed and/or inserted. A dry cleaning pad is at the end of the band and gently wipes away any deposits on the head.

Under normal use, the automatic head cleaner requires no special attention. However, on a heavily used VCR, the cleaning pad might become worn, and should be replaced. A worn pad does no actual cleaning, and under some circumstances, could pose a hazard to the VCR. Technically speaking, the cleaning mechanism is supposed to be replaced when the video head assembly is replaced. As video heads have an estimated life of some several thousand hours of use, you can usually go by for several years without worrying about replacing the automatic head cleaner in your VCR.

You should get into the habit of inspecting automatic cleaning mechanism each time you open the VCR for general maintenance. Check the pad to make sure it is not worn. If it is worn, replace it. Replacement cleaning pad parts are available from the manufacturer (though most dislike selling such parts to the general public) as well as from many of the VCR parts sources listed in appendix A. The exact replacement procedure differs between brands, but usually involves replacing the entire cleaning band.

NOTE: If you find the cleaning pad is severely worn or broken, it is best to remove completely the automatic cleaning assembly, if possible. This step will prevent the worn or broken pad from coming into contact with the delicate video heads. No automatic head cleaner is better than a broken automatic head cleaner.

### Cleaning costs

Prices for cleaning kits—either cassettes or swabs—vary. A pack of four to six swabs and a bottle of cleaning solution costs from $6 to $10. Cleaning cassettes range from $12 to $30. Either kit is the best investment you can make to keep your VCR alive and well.

## Oiling and lubricating

Videocassette recorders generally require little lubrication. After a year or two of use, however, it is advisable to apply small amounts of oil and grease to keep the deck in top-notch condition. At every major PM interval, you should lubricate such parts as the shaft of the capstan pinchroller, the threading mechanism, the loading mechanism, and the mechanical controls. Lubricate these parts more often if you use your deck outdoors where it is subject to extremes in temperature or excess dust and dirt.

You also should lubricate the mechanical parts if you clean the machine with a spray cleaner/degreaser. The cleaner acts to strip any oil or grease from the moving parts, and using the VCR without lubrication can wear it out prematurely.

Here are some good rules of thumb to remember when lubricating your VCR:

☐ If it spins, oil it
☐ If it slides or meshes, grease it

Light machine oil is the best lubricant for spinning parts, such as the bearings of idlers and rollers. You should never use a spray-on lubricant, such as WD-40. Go easy on the oil and apply it only to the rotating part. Never apply oil to the surface of a roller or where the lubricant might come into contact with the tape. Remember, more is not necessarily better. Excessive oil will gum up the tape causing clogged video heads, jammed tapes, and other maladies. Never apply oil directly to a motor. For small parts, apply the oil with a syringe applicator, as shown in Fig. 5-25 on the next page.

A nonpetroleum grease is the best lubricant for sliding and meshing parts, the bulk of which are found in the loading and threading assemblies and in mechanical controls (older decks only). This includes cams, bell cranks, gear surfaces, and levers. As with oiling, apply only a small dab of grease. Spread the grease by moving or sliding the part back and forth a few times.

If the part can't be easily moved or is geared down and won't budge, power the deck and insert a tape. To spread the grease, cycle the loading mechanism by inserting a cassette and using the front-panel controls of the deck.

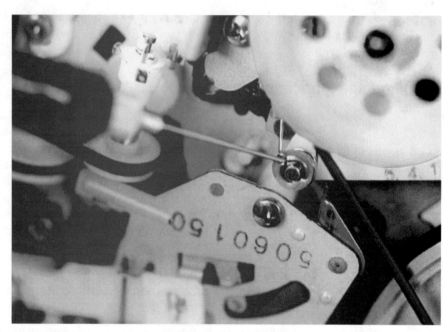

■ **5-25** *Apply a small drop of oil with a syringe or needle applicator to only those parts that absolutely require it.*

Use grease to lubricate the slides in the tape-threading mechanism. Apply a dab of grease to the end of the slide (see Fig. 5-26) and cycle the deck between the PLAY and STOP modes to spread the lubricant evenly (with Beta decks, repeatedly eject and reinsert the tape to activate the threading mechanism). You need to insert a cassette into the deck or the VCR won't respond. Use a cotton swab to wipe up excess grease. If the slide is dirty, clean it first before adding more grease.

Also apply grease to the mechanical controls on older VCRs (these machines have "piano-key" controls). Locate the levers and mechanisms for the PLAY, FAST FORWARD, REWIND, and other controls and apply a small amount of grease. Press the controls to spread the grease, and wipe up the excess.

Grease and oil also should be applied to the gears on the underside of the threading mechanism. In most VHS machines, the threading mechanism is driven by worm gears. These require grease to keep them from binding. As before, apply a small dab of grease and cycle the threading mechanism back and forth to distribute the lubricant.

Another subsystem of the VCR that needs lubrication is the cassette-lift mechanism. The cassette lift is on both top- and front-loading VCRs. Apply a small dab of oil to the shafts of the gears

■ **5-26** *Apply grease to parts that slide or mesh, such as the tape spindle slides in a VHS deck.*

and pulleys. Avoid using too much oil; it might find its way to the tape and cause problems later.

## Tuner cleaning

Mechanical detent tuners on early VCRs need regular cleaning (about every 24 to 48 months). If the tuner isn't excessively dirty, you can apply spray cleaner to it without removing any of the mechanism. Use a tuner cleaner/lubricant (available at Radio Shack) and squirt it liberally inside the tuner mechanism. Spin the tuning dials quickly in both directions to spread the cleaner.

If the tuner is very dirty, you might have to remove it from the machine to gain adequate access. Remove the front-panel knobs and locate the screws that hold the tuner in place. Remove them and gently lift the tuner out. The tuner will be connected to the VCR by several wires; if the deck uses connectors, unplug the wires and note where they belong. If the wires are permanently soldered in place, remove the tuner only enough to gain access to the internal contacts. Liberally spray the cleaner/lubricant, then replace the tuning mechanism.

Note that in the older VCRs, the tuner mechanism is composed of many channel strips—one strip for each VHF channel. Each strip tunes a specific channel. If the tuner is out of adjustment, the strips might need to be serviced or recalibrated. Although this requires some special tools and test tapes, it can be done at home if you follow the instructions in the service manual.

## Demagnetizing

Over time, the heads and other metal parts that come in contact with videotape will build up a small amount of residual magnetism. This magnetization can impair the signal quality of your tapes, resulting in a muddy or garbled sound and sometimes a weak picture. To return good sound to your tapes, regularly demagnetize the audio heads, the tape path, and all ferrous metal parts that come in contact with the tape.

Video heads also can be demagnetized, although most VCR manufacturers now recommend against it, saying it is unnecessary. You can do it as long as you use a demagnetizing tool specially made for use in VCRs and on video heads. Use of a regular reel-to-reel tape recorder demagnetizer can shatter the video heads, due to the intense magnetic field that is induced by the tool. If you decide to demagnetize the video heads in your VCR, proceed at your risk.

To use the demagnetizer, start with the tip about 3 feet from the deck. Turn on the demagnetizer, and slowly bring it close to the VCR. Wave the tip past the full-erase head, tape-guide spindle, audio heads, video heads, and other components. Do not physically touch any of these parts, or you can leave more residual magnetism than you had before. Be especially wary of touching the video heads with the tip of the demagnetizer. Even with the proper tool, actually touching the heads can shatter them.

When you are done, keep the demagnetizer on and slowly draw the tool away, again to a distance of about 3 feet. Turn the tool off. If power to the tool is turned on or off when the tip is closer than 1 or 2 feet from the deck, repeat the process. Otherwise, you will leave the deck magnetized.

Be aware that the demagnetizing tool can erase your cassettes. When using the tool, place all tapes away from the work bench.

## End-of-tape sensor cleaning

The end-of-tape sensor detects the clear leader portion spliced onto the beginning and end of all VHS and 8mm videotapes. An infrared (sometimes visible) light shines through the leader and strikes a photodetector, as shown in Fig. 5-27. When the photodetector sees light, it instructs the deck to stop playing (it also works during rewind and fast forward).

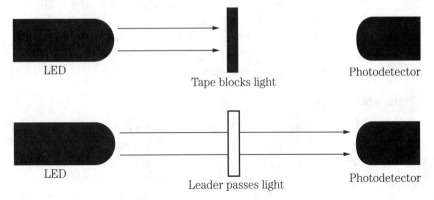

■ **5-27** *The operation of the end-of-tape sensor in a VHS recorder.*

If the sensor is dirty, it might never detect the light, and you lose the autostop safety feature. This malfunction can lead to stretched or broken tapes. Use a cotton swab dipped in alcohol to clean the sensors and the lamp assembly. In almost all VCRs, there are two sensors, located on either side of the loading/threading mechanism. Be sure to clean them both.

Beta VCRs use electromagnetic coils to sense the ends of the tape. Gently clean the coils with a brush and inspect the bracket that connects them to the deck.

## Reel spindles

The reel spindles rotate to supply and take up tape as it travels through the tape transport mechanism. Although you wouldn't know it, the action of the spindles is very delicate, and excessive dust or dirt can impair their performance, leading to considerable problems.

Use a soft brush to get rid of an accumulation of dust and grime. Apply a small amount of oil to the top of the spindle. Use only a small amount. Rotate the spindle to spread the oil on the shaft. Only if the spindle is exceptionally dirty or jammed should you remove it. The height of the spindle is precise and on most machines is determined by the number and thickness of small washers installed on the shaft. If you do disassemble the spindles, be sure to put them back together in the exact same sequence.

Note that under the take-up spindle there's a sensor. This sensor uses light to detect when the spindle is turning. If the spindle stops turning during playback (an event that can lead to tape spillage), the sensor relays this information to the VCR, and the VCR shuts itself off after 5 or 10 seconds. Because this sensor uses light, any dust, oil, or grime can impair its performance. While you have the take-up spindle off, use a brush to clean the sensor and the inside rim of the spindle.

## Transmission system

Many of the later models of VCRs—those made since 1990 or so—are equipped with a geared transmission system. This transmission system transfers power from a centralized motor (usually the same motor that drives the capstan) to the supply and take-up spindles. The transmission is most often self-contained, and is comprised of a series of plastic gears. After some period of use, the grease on the gears can become thin. You should apply a small amount of new grease to the meshing surfaces of the gears, and a dab (but no more!) of oil to the shafts of the gears.

Note: Be sure to lubricate the gears sparingly. A little bit of grease and oil goes a long way. Excess grease has a tendency of finding its way to your videotapes, which can ruin them.

There might be limited access to the transmission system from the top of the VCR. You might need to remove the bottom plate of the VCR to gain access to the transmission. If the transmission system is not readily accessible from the top or bottom, you should not disassemble the deck to try to get to it, unless you know the transmission is damaged and needs repair.

## Transmission clutch

A common part of the transmission system described in the previous section is a spring-loaded clutch. The clutch saves wear and tear on the entire VCR when the transmission "shifts." For example, if the VCR is in FAST FORWARD mode, and you press REWIND, the transmission shifts the drive power from the take-up reel to the supply reel. The clutch provides the necessary slack to prevent early mechanical failure and tape breakage.

On many models of VCRs, the transmission clutch is a simple device consisting of two plastic plates covered with a feltlike material. In time, the felt can be worn (it almost looks polished), and the clutch can slip when viewing a tape. The slip appears on the TV screen as a momentary glitch in the picture, and is often accompanied by a squeaking sound coming from the insides of the VCR.

You can often temporarily rejuvenate the clutch by spraying a very (accent on very) small amount of WD-40 on the clutch pad. Eventually, this trick will no longer work as the clutch continues to wear, and the clutch pad, the entire clutch mechanism, and sometimes the entire transmission system, has to be replaced. Fortunately, this doesn't usually happen until after many hundreds—and even thousands—of hours of use.

When greasing and oiling the transmission system, take special care that you do not get any on the clutch pad. Once the clutch pad has been contaminated with grease or oil, it usually must be replaced, which can be both costly and time consuming.

## Front-panel controls

The front-panel operating controls seldom need preventive care unless the VCR has been subjected to heavy doses of dust, dirt, damp air, and cigarette smoke. You can clean the bulk of the control switches with a brush or compressed air. The control panels in most decks use sealed membrane switches, so the internal switch contacts are not affected by dirt (the membrane switches are located behind the actual switch button, as shown in Fig. 5-28, which you push when operating the player).

If the switches in your VCR are the regular open-contact type, you might need to apply a small amount of electrical contact cleaner to dispel the dirt. Again, you need not perform this procedure unless

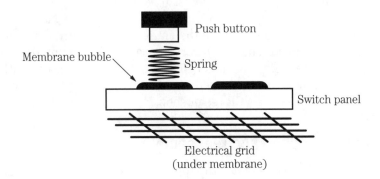

■ **5-28** *Membrane switches don't use exposed electrical contacts, so cleaning is not required. The contacts of the switches are located under or within the membrane bubble.*

the deck has been exposed to heavy amounts of contaminants or the controls are not working properly.

The contact cleaner dries quickly and without leaving a residue. Nevertheless, you should wait several minutes after cleaning before you turn the machine on and attempt to use it. As usual, never clean the switch contacts when the machine is turned on or plugged into the wall socket.

### Jog/shuttle control

Many high-end VCRs come equipped with a jog/shuttle control. The jog control is a round knob that you spin to advance the tape one frame at a time. The shuttle control, placed around the jog control is a ring that operates the various fast forward and reverse special-effects modes. The amount you turn the shuttle control in either direction determines the speed of the fast forward and reverse special effects.

The jog/shuttle control is composed of many electrical contacts (this applies to the jog/shuttle control on the VCR as well as the remote control, if it also is equipped with a jog/shuttle pair). These contacts can become dirty after extended use of the VCR, and might require periodic cleaning. You can clean the jog/shuttle control with regular contact cleaner. Make sure the cleaner does not contain a lubricant.

In some cases, when the jog/shuttle controls are used heavily or when they are abused, the contacts can become worn out. There is little that can be done to correct this, apart from replacing the jog/shuttle control mechanism, which can be expensive. For the most part, if you find yourself twirling the jog/shuttle control quite a bit, you should use the jog/shuttle pair on the remote control as much as possible, rather than on the VCR. If the jog/shuttle control on the remote control wears out, it's a simple and relatively inexpensive job to replace the entire remote control (the cost of a replacement remote control with jog/shuttle is about $100, depending on the make and model). The cost to replace the jog/shuttle control on the VCR is considerably higher, especially if you decide to take the deck to a VCR repair shop.

## Final inspection and reassembly

Before you replace the top and bottom covers, double check your work. Make absolutely certain you haven't left tools, hardware, used cotton swabs, or cleaning supplies inside the chassis of the VCR. Check to make sure that you have properly replaced all parts.

If everything looks satisfactory, replace the covers and reassemble the deck with the screws you removed. Be sure the covers are aligned properly before tightening any of the screws. When all the screws are in place, tighten them lightly. Do not overtighten or you might strip the threads of the screw and chassis.

## Checkout

After every PM interval, check the deck for proper operation. You should use one known-good tape (either a commercial tape or one you've made—see chapter 4 for details). Keep these operational points in mind during the checkout process:

☐ The tape-loading mechanism operates smoothly and accepts the tape as it should
☐ The tape threads and unthreads without mishap (try the sequence several times)
☐ The tape plays properly, with full audio and video
☐ The FAST FORWARD and REWIND buttons operate properly and shuttle the tape to the end
☐ The deck automatically shuts off when it reaches the end of the tape (and the same function should work when rewinding the tape)
☐ Operating the tuner selects the different channels
☐ All front-panel indicators operate normally
☐ All the function buttons on the front panel of the deck and the remote control work as they should

## Maintenance log

You might want to keep a logbook of all the maintenance checks and cleaning you've performed on your VCR. Use the sample log form that appears at the end of this chapter, or make your own. In your log, be sure to note the exact checks and maintenance procedures you performed and when you performed them. If you have replaced any parts (like tires and belts), note their part numbers in the log and troubles you might have had when installing them. By keeping a thorough log, you can better estimate the timing of the preventive maintenance intervals for your specific deck.

## Remote cleaning

Almost all VCRs are equipped with wireless infrared remote control units. These allow you to operate the deck from afar—play a tape, scan through a tape, and more—all from the comfort of your easy chair. Because the remote control can be taken anywhere in the house, it's often subjected to a lot more abuse than the other components in your video system. You might pride yourself in a clean VCR, but the remote could be filled with dirt, grime, or sticky syrup from a spilled soda pop.

If your VCR has a wired remote, you should follow the same general cleaning procedures. Also note that you should inspect and repair any damage that might have occurred to the wire leading between the remote and VCR.

### Exterior cleaning

Cleaning the remote control is straightforward and easy. Start with applying a damp cloth to the outside and remove exterior dirt and grime.

### Battery-compartment cleaning

Remote controllers are battery operated. Given the right circumstances, all batteries can leak, and the spilled acid can corrode the electrical contacts and can impair proper operation of the controller. If your remote control is operating sporadically, this might be the problem.

If the batteries have leaked or if the battery contacts have become only slightly dirty, clean them as follows. You also can use this procedure for cleaning the battery compartment in portable VCRs and camcorders.

☐ Carefully remove the batteries and discard them. Immediately wash your hands to remove any acid residue. Battery acid burns.
☐ Wipe off the battery acid residue with a damp cloth, and remove as much of the excess from the contacts as you can.
☐ As shown in Fig. 5-29, remove the remaining residue by rubbing the contacts with the tip of a pencil eraser. Blow the eraser dust out of the battery compartment when you are through.

Test the remote with new batteries. If the battery leakage is more extensive, you will have to take more drastic action. Refer to chapter 6 for more details on how to clean heavy deposits of battery acid.

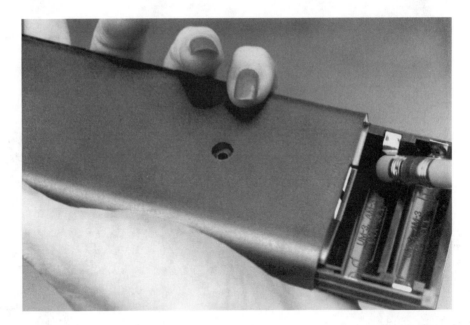

■ **5-29** *Use a pencil eraser to clean the battery contacts.*

## Interior cleaning

To clean the inside of the remote, you must disassemble it by re-moving the screws holding the two halves together. The screws are small and you will need a set of jeweler's screwdrivers to re-move them properly. Some screws could be hidden in the battery compartment, under the manufacturers label, or even beneath the thin metal cover on the front of the unit.

When the screws have been removed, separate the two halves. Clean the inside of the controller with a soft brush or can of com-pressed air. If liquids have been spilled inside the controller, spray the inside of the controller with a suitable cleaner/degreaser.

With most remote controllers, the circuit board is attached to the front panel and must be removed if you want to access the front switches. With the remote resting on a flat surface, remove the cir-cuit board screws carefully and gently lift off the control board. The plastic push buttons might be loose, so be sure not to knock them out of place when you remove the board (if so, putting them back isn't hard, but it can be time consuming). Clean the switches and front side of the circuit board with a spray from the can of cleaner/degreaser, as shown in Fig. 5-30 on the next page.

■ **5-30** *You can use spray cleaner on the switches in the remote control.*

## Remote control testing

Assume you've replaced the battery in the remote controller and you've cleaned it inside and out. It still doesn't seem to work. Because the infrared light from the remote is invisible, it's difficult to test that it is indeed working or if the problem lies in the VCR.

Many remote controllers have visible LED indicators that flash when you push a button on the transmitter. The LED is typically connected with the high-output infrared LEDs that send the remote control information to the player. You can be fairly certain that the controller is operating properly if the visible LED flashes.

Like everything else, the infrared LEDs in your remote controller do not last forever. It's possible, but not really likely, that one or all of the high-output infrared LEDs have gone out. If only one of the several LEDs that are typically built into the controller is faulty, you should still be able to operate the remote, but you might have to get in close to the deck.

One way to accurately test the remote is to use the infrared sensor described in chapter 4. This sensor is sensitive to infrared light and will register the light emitted by the remote controller. To use the sensor, position it closely to the front of the controller and start pushing buttons. If the LED on the sensor lights up, the remote is good. The fault probably lies in the deck. If it does not light up, double-check the operation of the sensor as described in the last chapter and try again. If the LED on the sensor still doesn't light up, you can be fairly certain the remote control is at fault.

Lacking the infrared sensor, you can use an ordinary AM transistor radio to test the remote. Dial the radio to 530 on the dial (or thereabouts if there is a station broadcasting on that frequency) and place it near the front of the remote. Push the buttons—you should hear a chirping sound if the remote is working. Note that this test does not actually verify light output from the remote but that the electronics in the remote are working.

Be aware that there are a variety of reasons why a remote control does not operate properly, and none of them have anything to do with a faulty controller or sensor in the VCR. If you are having trouble with your remote control and have performed the routine cleaning outlined here, check the remote-control troubleshooting charts in chapter 9 for more information.

## Tape troubles

Sometimes sound and picture ailments are not the fault of the VCR, but the videotape. Here's how to make sure:

### Old tapes

Be on the lookout for troubles arising from aging, worn-out tapes. Older tapes have a tendency to shed their magnetic oxide coating. The oxide can cause dropout (the small white flecks that dash randomly across the screen). Circuitry inside the VCR compensates for some of the dropout, but a badly deteriorated tape can be unwatchable.

If one of your tapes gets so bad that it's difficult to view, follow these steps.

☐ Make sure it's not the VCR's fault by testing the machine with a known-good tape.
☐ If it is indeed the tape that's bad, as a second step, cover the cassette with a white sheet, say a few nice words, and chuck it.

### Stretched tapes

Old or mishandled tapes can stretch, which causes the tape to curl. This in turn creates several problems, one of which is a bending or hooking of the picture at the top of the screen. Badly stretched and curled tapes can bring about other malfunctions, such as poor audio and a loss of picture synchronization. Unfortunately, like old and shedding tapes, little can be done to help "unstretch" a tape.

## Jammed tapes

VCRs that have not been maintained have been known to "eat" tapes. During recording or playback or during threading, the tape jams within a VCR. If a jam occurs, immediately stop the VCR and press the EJECT button.

The tape might be tangled; if it is, turn off the deck and remove the top cover. Manually extract the tape, being careful not to touch it. If you have a pair of clean cotton gloves, now is the time to wear them. If you don't have gloves, handle the tape by the edges only, because oil from your skin can cause head clogging and further tape damage.

When the tape is clear from all obstructions, take the cassette out of the loading port, if you haven't done so already. If the cassette cover closes on the tape, you'll need to reopen it by inserting a flat-blade screwdriver into the cover release button, as shown in Figs. 5-31 A and B.

Both VHS and Beta tapes have locks to prevent the reel tables from turning when the tape is removed from the machine. Opening the door on the Beta tape automatically disengages the reel

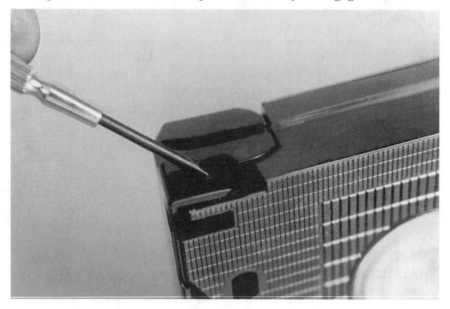

**A**

■ **5-31** *The location of the cover release pin on Beta cassettes (A).*

**B**

■ **5-31** *The location of the cover release pin on VHS cassettes (B).*

locks. VHS tapes have a small button on the bottom (Fig. 5-32) that must be depressed to release the locks (8mm tapes have a similar button on the bottom). Do not attempt to force the reels to turn—you'll snap the tape in two.

If the tape is severely damaged, it might be permanently ruined. You can cut out the bad portion and splice the tape back together, but this should be done at the beginning of the tape only. See the next section for more information on tape splicing.

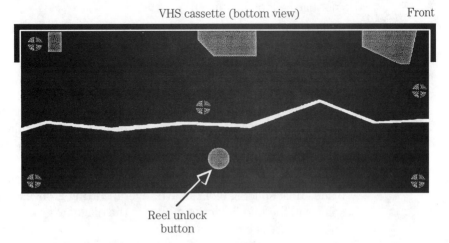

VHS cassette (bottom view)                      Front

Reel unlock
button

■ **5-32** *The reel unlock button on VHS cassettes is located on the bottom, inside a large hole.*

### "Scalloped" tapes

When the transport mechanism of a VCR is not aligned properly, it can cause the tape to ride too high or too low through the tape path inside the deck. This invariably causes damage to the edges of the tape. The most common cause of damage is tape scalloping, so-called because when viewed in the light, the tape appears to have a rippled, or *scalloped*, look.

Most VCRs can compensate for a small amount of scalloping, depending on whether the audio track or control tape edge of the tape has been damaged, and the extent of the damage.

☐ If the VCR is a hi-fi model, and if the scalloping is on the audio track edge of the tape, the tape will be viewable if you listen to only the hi-fi audio tracks.
☐ If the scalloping is on the control-track edge of the tape, the VCR might not be able to track the tape adequately if the scalloping is moderate or severe.

### Splices

Unlike audio tape, it is not advisable to splice videotape except for the clear leader portion at the beginning. Splices along the length of the tape aren't recommended because they can cause damage to the spinning video heads. Since the leader on videotape never passes by the heads, it is permissible to splice the tape in this area.

To splice videotape, gather the necessary tools around you: a pair of cotton gloves, a splicing block for ½-inch tape (or 8mm tape if you are splicing an 8mm cassette), a razor blade, and splicing tape (do not use cellophane tape).

☐ Pull the tape out of the cassette until all of the damaged parts are exposed. Keep the cassette door open with a piece of masking tape or a rubber band. Be sure to unlock the reels when you are pulling the tape out or you might stretch it.
☐ Cut out the bad portions and throw the tape away.
☐ Lay the two ends of the tape on the splicing block as shown in Fig. 5-33. Lay the tape so the shiny side is up (if both sides appear shiny, orient the tape so the surface on the inside of the wound reels is facing you).
☐ Overlap the tape slightly and cut both strands with the razor blade. Most splicing blocks have several cutting grooves: for splices in the middle of the tape, use the diagonal groove; for end cuts, use the diagonal one.
☐ Immediately splice the two ends with tape. Be sure that the tape ends don't move. Avoid gaps or overlap in the splice, as shown in Fig. 5-34.

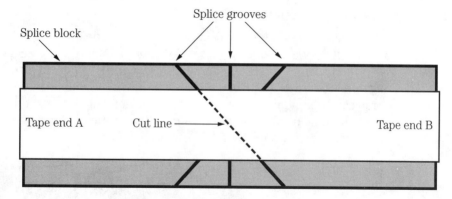

Cut tape ends with razor blade
over one of the cut lines.

■ **5-33** *Use one of the diagonal grooves in the splicing block to trim the tape ends.*

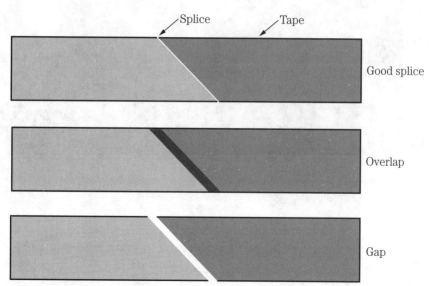

■ **5-34** *Good splices do not have gaps and overlaps.*

## Tape shell damage

The cassette cartridge itself (called the shell) might become damaged for a number of reasons: it could get crushed when stored in the bookshelf, broken in transit through the mail system, snapped apart when packed in the suitcase of your latest worldwide tour, or warped by the heat of the sun. You can sometimes repair a cracked shell by using modeler's cement. If the shell is beyond repair but you want to save the tape that's inside, you can transfer the tape from one cassette body to another.

### Using modeler's cement

Small cracks can be fixed by applying liquid modeler's cement to the shell. Use a small "00" or smaller brush to apply the cement. Dip the brush in the bottle and lightly touch the tip of the brush against the crack. Capillary action will draw the cement into the crack and spread it. Do not apply too much cement or it might drip onto the tape inside. You also can use Super Glue or other strong adhesive, as shown in Fig. 5-35.

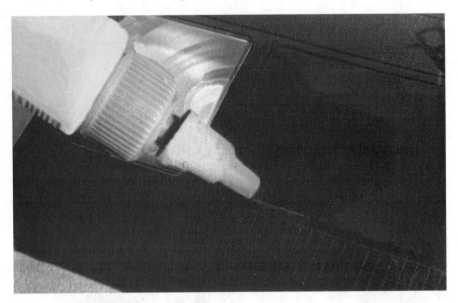

■ **5-35** *Apply a small amount of solvent-based cement to repair cracks and breaks in plastic pieces and tape cassettes.*

### Transferring tape to another shell

To replace the shell, you need an extra tape that you don't care about losing. A bargain basement tape recently bought at the five-and-dime store or a worn-out tape can be used. All you are interested in is the shell, so the condition of the tape itself is of no consequence.

Use a demagnetized screwdriver to loosen the screws on both cassettes. Open the shells on a flat surface, like a large table. Be sure that none of the internal parts, like rollers and springs, fall out. If they do, quickly note their position and replace them. When disassembling an 8mm cassette, use a jeweler's screwdriver.

Examine the way the tape threads around the rollers, posts, and other parts of the shell. Make a diagram if you think you might forget. Figure 5-36 shows a VHS tape opened, showing the threading course.

■ **5-36** *A VHS cassette, opened to show the supply and take-up reels.*

Take the two reels of tape from the good shell and gently lift them out. Place the tape aside or throw it away. Now, lift the two good reels out of the bad shell and place them in the good shell. Carefully thread the tape around the posts and rollers, and reassemble the shell. When replacing the screws, do not overtighten. With most shells, the screws are self-tapping and you'll strip out the plastic if you exert too much pressure.

Test the transplant by playing the tape on your VCR. Listen for squeaks or odd rubbing sounds—tell-tale indicators that the tape is not properly threaded within the shell. If so, take the shell apart and try again.

## Tape care

Follow these simple guidelines to maximize the health and longevity of your videotapes:

- [ ] Store all tapes in an upright position (see chapter 3 for more details).
- [ ] Store tapes in a cool, dry place away from direct sunlight.
- [ ] Avoid placing tapes on the floor or carpet where they might pick up dust and fibers.
- [ ] Never touch the tape surface; wear gloves when handling or splicing tapes.
- [ ] Place tapes several feet from electromagnetic sources such as loudspeakers, TV sets, and telephones.

# Maintenance Log

Videocassette recorder maintenance and repair log

## VCR vital statistics

Brand _____
Model _____
Date purchased _____ Serial number _____
Where purchased _____
Salesperson _____Warranty period _____
New_____Used _____ If used, how old when purchased? _____
Type: ac operated home _____ Portable_____ Camcorder_____

## VCR design

Number of video heads: 2 _____ 3 _____ 4 _____ 5 _____
Format: VHS _____ Beta _____ 8mm _____ Other _____
Audio: Stereo _____ Hi-Fi _____ MTS ready ___ MTS Adapat ___ Dub ___
Special circuits: Super Beta ___Super VHS ___HQ ___Digital effects ___

## PM schedule supplies (type or brand and source)

Household spray cleaner _____
Nonpetroleum solvent cleaner _____
Optical lens cleaner_____
Oil lubricant_____
Grease lubricant _____
Electrical contact cleaner_____
Compressed air_____
Rubber cleaner_____
Other: _____
_____
_____
_____
_____
_____
_____
_____

## Tools (List)

_____
_____
_____

General maintenance

| | | | | |
|---|---|---|---|---|
| Date | | | | |
| Clean PCB | | | | |
| Clean heads | | | | |
| Clean transport | | | | |
| Clean controls | | | | |
| Check operation | | | | |
| Inspect wiring | | | | |

Lubrication

| | | | | |
|---|---|---|---|---|
| Date | | | | |
| Loading assembly | | | | |
| Threading assembly | | | | |
| Transport assembly | | | | |
| Other _____ | | | | |

Contact cleaning

| | | | | |
|---|---|---|---|---|
| Date | | | | |
| Power switch | | | | |
| Front panel switches | | | | |
| Interlock switches | | | | |
| Remote switches | | | | |
| Remote battery terminals | | | | |
| Connectors | | | | |
| Head assembly | | | | |

179

*Tape troubles*

Page 3 Videocassette recorder maintenance and repair log Page 3

Repair/replacement

| | | | | |
|---|---|---|---|---|
| Date | | | | |
| Capstan belt | | | | |
| Idler tire | | | | |
| Idler drive belt | | | | |
| End-of-tape sensor/lamp | | | | |
| Switch | | | | |
| Capstan pinchroller | | | | |
| Capstan motor | | | | |
| Main PCB | | | | |
| Video head assembly | | | | |

Problem occurrence

Date ____________
Description of problem ______________________________
______________________________
______________________________
Action taken ______________________________
______________________________
______________________________
Meter readings, measurements, etc. ______________________________
______________________________

Date ____________
Description of problem ______________________________
______________________________
______________________________
Action taken ______________________________
______________________________
______________________________
Meter readings, measurements, etc. ______________________________
______________________________

Problem occurrence (continued)

Date _____
Description of problem _____
_____
_____
Action taken _____
_____
_____
Meter readings, measurements, etc. _____
_____

Date _____
Description of problem _____
_____
_____
Action taken _____
_____
_____
Meter readings, measurements, etc. _____
_____

181

Date _____
Description of problem _____
_____
_____
Action taken _____
_____
_____
Meter readings, measurements, etc. _____
_____

Special notes

_____
_____
_____
_____
_____

# VCR first aid

ACCIDENTS DO HAPPEN TO EVEN THE MOST CAREFUL PEOPLE and to their VCRs. If an accident befalls your videocassette recorder—you spill your cup of coffee into it, for example—there is no need to panic. By following some simple cleaning and checkout procedures, you can reduce or eliminate potentially costly repairs.

This chapter discusses first-aid treatment for a number of accidents and how you can repair all or nearly all of the damage yourself. Even if your VCR is presently healthy and working fine, read this chapter to acquaint yourself with the recommended procedures. You never know when they'll come in handy.

## Tools and safety

Most of the repair procedures require disassembly of the deck. If you haven't done so already, please read chapters 4 and 5 for more information on how to take your VCR apart and what tools you need to do the job right. Pay particular attention to the safety points in chapter 5.

Keep in mind that the steps outlined in this chapter are designed to help you minimize the damage of accidents, not make them worse. If you are not confident in your ability to perform some procedures or lack the proper tools and cleaning supplies, then by all means don't do them. When in doubt, refer to a qualified technician.

## Dropped deck

VCRs and camcorders are not meant to be dropped off book-shelves, TV racks, tables, or car seats, but it happens. A drop onto soft carpeting might just shake up the deck a bit but not cause serious internal injury. A harder impact, however, can break or bend the cabinet or chip off a piece of the front panel.

A cracked cabinet or other broken piece on the exterior of the machine can usually be mended with a strong adhesive. If the part is plastic, almost any clear glue recommended for use on plastic should suffice. After gluing, hold the pieces together with your fingers until the adhesive sets. Or, tape the pieces together like a surgical suture until the glue is completely dried.

## Checking for internal damage

After a VCR has been dropped, even if there is no visible damage, inspect it carefully. Don't test it by plugging it in and playing a tape. You might cause additional damage.

You can quickly check if parts have come loose inside the deck by unplugging the unit from the wall socket and gently tilting or shaking it. If you hear rattles, you can bet something is bouncing around that shouldn't. Even if you don't hear loose parts, it's a good idea to disassemble the VCR before attempting to use it. Follow the general disassembly instructions provided in chapter 5.

When the top cover of the cabinet is removed, look for the loose parts and find out where they came from. If the broken piece is plastic and is not an extremely critical part, glue it back on with a suitable cement. A solvent-based cement can be found at hobby or plastics supply stores. Broken metal parts are harder to mend, but there are a number of adhesives for metal that might do the job. The alignment of internal parts is often critical to the operation of the VCR, so make sure you glue the repaired piece on straight.

While the top is off, inspect the interior of the deck for hidden damage—things other than completely broken pieces. Pay particular attention to circuit board(s), the tape-loading mechanism, the tape-threading mechanism, and the video head drum. If any part of the works looks broken, bent, or out of place, return the deck to a qualified repair center. A broken head cannot be serviced; it must be replaced, and replacement requires exact alignment using special tools.

Check the printed circuit boards inside the VCR. If you see any hairline cracks, it's a bad sign. A broken printed circuit board must be replaced. Using the deck as it is can cause more damage because components might be shorted. You could conceivably burn out the power supply, motors, and other costly parts by attempting to operate the VCR with a faulty circuit board. Fortunately, however, it takes a very healthy jolt to break a circuit board, so it is not a common occurrence.

## Broken wires

A strong impact can cause electrical wires and connectors to break or come off. Sometimes the connector, like the one in Fig. 6-1, is just jostled in its socket. Look very carefully at these because some connectors might look fine from the outside but electrical contact inside has been broken. Press all the connectors firmly into their sockets just in case. If the connector has come out all the way, just plug it back in to its respective socket.

■ **6-1** *A ribbon cable and connector attached to a circuit board.*

Broken wires leading to connectors and boards must be resoldered. If the solder joint is accessible and does not require precision work, you can do the work yourself. If you are unfamiliar with proper soldering techniques, refer to appendix C for a quick lesson.

If more than two wires have come undone and it is not obvious where they go, consult the schematic diagram for your VCR, if available, or return the deck to a service shop. Don't take chances here. If two wires are loose, you have a 50 percent chance of soldering them back properly. You also have a 50 percent chance of soldering them back in the wrong place—not particularly great odds, considering the permanent damage that can ensue.

## Leakage current test

Prior to turning on your VCR, even if you do not take it apart to inspect for internal damage, be sure to check for possible leakage current. This test determines if any part of the ac line has come in contact with the metal cabinet or base. It is a safety check to prevent a potential shock hazard. You'll need a volt-ohmmeter to perform the test.

### Method 1

With the VCR unplugged, short the two flat prongs on the end of the ac cord with a jumper wire, as illustrated in Fig. 6-2. On the meter, select the ohm function and a range of no less than 1 k$\Omega$. Connect one test lead from the meter to the jumper on the ac cord. Connect the other test lead to any and all bare (not painted) metal parts of the deck. For a typical VCR, the meter should read about 50 to 100 k$\Omega$—if you get any reading at all.

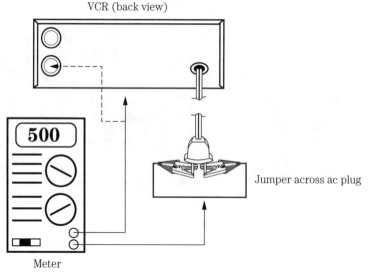

■ **6-2** *One method for testing leakage current with a volt-ohmmeter.*

A reading substantially lower than this is a good indication that the power supply has come into contact with the metal parts of the deck. If this happens, inspect the power cord as it leads into the deck, as well as the ON/OFF switch, the transformer (usually bolted onto the back of the machine), and the wires leading from the transformer to the printed circuit board. If the wires are broken or are shorted to the cabinet, repair the fault before using the deck.

Note that not all exposed metal parts of the VCR will return a high value. Touching the center conductor of one of the audio output connectors could yield a very low resistance, say from 1 to 50 Ω. This resistance is normal with some machines.

## Method 2

Follow the circuit diagram shown in Fig. 6-3. Place the meter in the ac millivolts range and take a reading. The value should be very low—a matter of 2 to 10 mV (millivolts). Anything higher might indicate a partial or complete short circuit. Note that you need to connect one side of the meter to a suitable earth ground. The ground connector in a wall outlet is a good choice. If your cable TV system is adequately grounded (the outer conductor of the coaxial cable is attached to a grounding rod), you can use it instead.

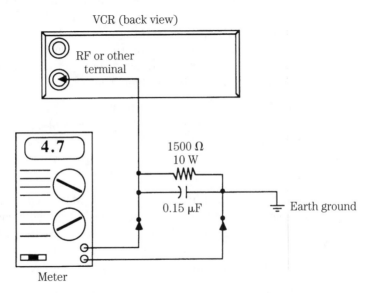

187

■ **6-3** *Alternate method for testing leakage current. This method is preferred because it gives an actual reading of the current.*

# Fire damage

Fire damage includes the effects of both the heat and the smoke of fire. Intense heat can destroy the VCR, of course, but even moderate heat (150°F to 175°F) for a short period of time can cause considerable damage to the deck's electrical and mechanical components. After a fire, check for obvious damage to plastic parts.

If a tape was in the deck, remove it and inspect it for warping, melting, or other damage. If it is warped, then there is a good indication that the other plastic parts inside the deck are warped as well. Disassemble the machine and inspect it thoroughly.

Minor warping of plastic parts, as well as expansion of metal parts during the heat of a fire can throw components out of alignment. You can test this by playing a tape from beginning to end (you can fast forward through the tape to save time). If anything is out of alignment, such as the tape-threading mechanism, playback will be impaired for at least some portions of the tape. Perform this test only after thoroughly inspecting the deck and are certain that powering it up won't cause additional damage.

## Smoke damage

Many people erroneously think that heat from a fire causes the greatest damage. Insurance companies will tell you that smoke, not heat, is the worse enemy. Why? Smoke gets everywhere, even in rooms that were not touched by the fire itself. The black soot from smoke cakes on everything and has an acidlike effect that can eat through finishes and protective layers. If your VCR works at all after a fire, failure to remove the layer of soot in its interior could cause considerable problems later.

If your VCR survives a fire, clean it off (and any other electronic gadgets for that matter) as quickly as possible. Immediate cleaning minimizes smoke damage to the exterior cabinet. Thoroughly wipe the outside of the VCR using a damp sponge to pick up the bulk of the soot.

Matters are a little different on the inside of the deck. Using a household cleaning spray on the mechanical and electronic parts of a VCR is not recommended, partly because the cleaner is water based and poses a potential short circuit hazard, and partly because it is not really effective in cleaning hard-to-reach places.

A cleaner/degreaser spray can be used to clean the inside of the deck thoroughly. Spray the cleaner heavily to remove the soot. Use the extension spray tube (usually included with the can) to get at hard-to-reach places. The cleaner evaporates after 5 or 10 seconds. If soot remains, spray again. If the smoke buildup is heavy, spray the cleaner on and brush it off with a small painter's brush.

The cleaner/degreaser leaves no residue, but it can remove the lubrication on some mechanical parts in the deck (including the threading posts and rotating parts). After using the cleaner/degreaser, lubricate the deck, if necessary, following the instructions provided in chapter 5. But remember that VCRs, on the whole, require little lubrication.

If you're not sure if a part needs oiling or greasing, try using the deck for a while. Watch its operation carefully to see if lubrication is called for (leave the top cover of the VCR off to do this).

Finally, the heads, including the rotating video heads, will need a complete cleaning as well. Follow the cleaning procedures discussed in chapter 5.

# Water damage

Each year, water does more damage to personal property than fire. Even if you don't live in a flood basin, there is always a chance that your VCR can be subjected to a liquid drink. If you have a camcorder, you could drop it into the water while lounging near the pool. It's happened before, and it can happen again.

If your VCR is ac operated and it becomes wet, immediately unplug it. If it is unsafe to reach the plug, turn the power off at the circuit breaker or fuse box. Do not touch a wet VCR; you might receive a bad shock. If the deck shorts out when water spreads inside its cabinet, there is a very good chance that it is seriously damaged and should be taken to a repair shop for proper service. If there is no sign of immediate short circuit damage, you can minimize any further problems by completing the steps that follow.

## Removing excess liquid

Of first importance is to soak up the excess liquid. Use paper towels to blot up the extra. If you feel any liquid has seeped into the machine, you must disassemble it and remove the standing water from the inside as well.

If the deck was dunked into fresh water, you need only to wait until the remaining moisture evaporates. Some water could be trapped under components, so even if the surface of the circuit board and internal parts are dry, there still might be water lurking underneath. You might use a hair dryer—on low or no heat only—to help speed up the drying process. Do not dry with high heat, as you might warp some parts. Even after the water is gone, some moisture might still be present, particularly on smooth metal parts and around the head drum area. Allow another two to three hours for condensation to evaporate.

You can test for remaining moisture by placing the VCR in a plastic garbage bag. Seal the end and place the bag overnight in a warm but dry place. If there is moisture remaining in the deck, condensation will form inside the bag. It's a clue to let the machine dry out some more.

Some decks have a built-in dew circuit that prevents playback if there is condensation in the unit (older decks had a dew light that turned on to indicate excessive moisture). If your VCR still doesn't work after drying, condensation might be the problem. Pay particular attention to the head drum. Condensation on the polished surfaces can cause the tape to stick and jam or stop and cause further damage. If the deck turns on but won't play a tape, keep it on for a while to help burn away the moisture.

### Removing sticky or staining liquids

If the VCR was subjected to sticky or staining liquids—such as saltwater, coffee, soda pop, or milk—you need to clean the deck thoroughly inside and out. Cleaning prevents residue from the liquid from interfering with the operation of the unit (saltwater will corrode the metal parts, for example). If the liquid is thick, use a damp sponge to wipe up as much as possible.

Next, thoroughly spray the VCR with a cleaner/degreaser, as mentioned in the previous section on fire damage. Apply the cleaner/degreaser heavily until all signs of the residue are gone. Because the cleaner/degreaser is not water based, you also can use it to remove any water that might remain in the deck. The cleaner/degreaser acts to displace the water, removing it from even hard-to-reach places.

The remote-control transmitter is often subjected to heavy abuse because it can be placed on tables with food and drinks. Crumbs, liquids, and other matter can easily fall in the cracks and muck up the switch contacts inside or short out wire terminals and connections. Disassemble the remote, as explained in chapter 5, and thoroughly clean the switches. You can spray the cleaner on or use a damp cotton swab, as shown in Fig. 6-4.

After cleaning, moving parts might require lubrication. See chapter 5 for more details on lubricating VCRs. Be sure to clean all the magnetic heads prior to use.

### After-cleaning test

When you are satisfied that all the moisture has been removed and that the deck has been properly lubricated and cleaned, perform a leakage current test as explained earlier in this chapter. If the leakage test proves negative, plug in the unit and turn it on. Test the VCR completely for proper operation. Should you find that

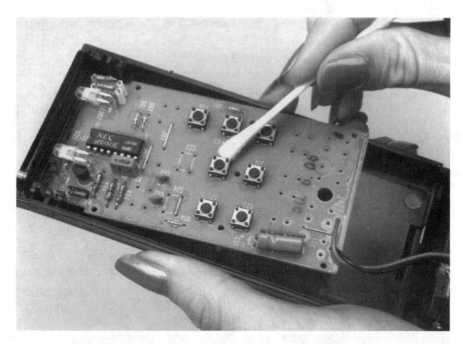

■ **6-4** *Clean the contacts and switches in the remote control with a cotton or sponge swab dipped in a suitable cleaner.*

some moisture is still present inside the VCR, stop testing immediately and allow the deck to dry out more.

## Sand, dirt, and dust

The video heads in a VCR are extremely sensitive to even a small amount of dirt. Decks used outside in dusty or sandy environments are susceptible to premature failure, especially if they are accidentally dropped in the sand at the beach or carried on a long-distance desert hike, where the dust from the great outdoors permeates everywhere.

Removal of every last speck of sand, dirt, and dust is crucial to ensure proper operation of the VCR and to prevent possible damage to the video heads, video-head drum, and tapes. Thorough cleaning is particularly important if the deck is filled with gritty sand. A soft brush can be used to wipe away the bulk of the sand, or if there is lots of it, you can use a vacuum cleaner with a brush attachment. Remaining sand, dirt, and dust can be removed from the cabinet, chassis, and interior with a tack cloth designed for wiping away sawdust from woodworking projects. Tack cloths are available at any hardware store.

Even though the visible dirt and sand might be gone, there still could be an accumulation of microsized particles of dust. You can clean the entire inside of the VCR by liberally squirting it with your trusty can of cleaner/degreaser. The fluid in this cleaner evaporates quickly, leaving no residue. The cleaner/degreaser is a solvent, but it is not petroleum based so it will not harm metal or rubber parts.

Sand and dirt have a way of getting into the most unusual places, and it might take several hours to displace it all. Be sure to inspect the input and output jacks on the deck. Grit inside prevents proper electrical contact, so you might not get any sound or picture. You can clean the jacks with a swab soaked in alcohol (see Fig. 6-5). Clean both the inside and outside.

As mentioned before, you might need to oil and lubricate the spinning and sliding parts in the deck after cleaning. Follow the directions outlined in chapters 4 and 5, and be sure to go lightly on the oil and grease.

■ **6-5** *The input and output terminals can be cleaned using a cotton swab dipped in alcohol or nonpetroleum solvent cleaner.*

## Foreign objects

Almost any service technician of audio and video equipment will tell you that a good portion of machine repairs are caused by people inserting foreign objects into the mechanism. Horror stories of this sort abound, and most are innocent mistakes caused by children. How about the one where a five-year-old boy stuck a peanut butter sandwich into the tape loading drawer because he wanted to know what type of picture it would make? Or when a young girl used her father's VCR as a piggy bank? She put all her extra pennies, nickels, dimes, and quarters into the ventilation slots of the deck. She had amassed quite a savings until the VCR suddenly went on the fritz.

This type of accident is best avoided by warning young children that the VCR is not a toy. You can make your own rules in your home, but if you allow your children to use or even touch the deck, spend a few moments instructing them on the right and wrong way to play tapes. Children are naturally curious about things, especially something new like a VCR, and you'll reduce the chance of a serious accident if you provide adequate do's and don'ts ahead of time.

Despite the best efforts, warnings, and rules on your part, foreign objects might still get lodged inside the machine. Some might even be your fault. When this happens, immediately turn off the deck and unplug it. If the object cannot be completely retrieved, disassemble the VCR and remove it.

## Leaked batteries

Portable VCRs and camcorders, as well as remote controls for home decks, run on battery power. Given the right set of circumstances, even the best-made battery can leak. The acid from the battery oozes everywhere and not only blocks electric current, causing failure, but corrodes the inside of the battery compartment and the battery terminals.

Your best defense against leaking batteries is to remove them if your deck or remote will not be used for a long time. Batteries tend to leak the most when they run out of electrical juice, so a battery that sits unused in a portable or remote has a very good chance of leaking. If the batteries are still good when you take them out (you can test them with a battery tester or volt-ohmmeter), use them in something else.

When storing batteries—new or used—keep them in a cool, dry place. You can prolong the shelf life of batteries by storing them in the refrigerator (not freezer). Again, batteries can leak about everywhere for any reason, so to prevent contaminating your food with battery acid, wrap them up in a sealable food storage bag.

## Removing battery acid

Should the batteries leak in your camcorder, portable, or remote, remove them immediately and throw them away. Avoid excessive contact with battery acid, as it can burn skin. Use a lightly damp-ened cloth to remove the excess battery acid deposits from the battery compartment. If the batteries leaked only a little bit, there might not be any deposits in the compartment or on the terminals. If the terminals look clean and bright, install a new set of batteries and test the unit.

If there are excessive battery-acid deposits, clean the entire com-partment with isopropyl alcohol or with the can of cleaner/de-greaser. Use a cotton swab to scrub the cleaner onto the terminals. Also, you can remove any remaining battery acid deposits by rub-bing the surface of the contacts with a pencil eraser.

# Broken remote

On most modern VCRs the remote control is the lifeline to the VCR. In fact, without the remote control, most of the advanced functions of the VCR—such as setting the clock and programming the deck to automatically record shows—are unavailable. If the re-mote control is not working properly you might be able to play tapes, but little more.

The most common cause of broken remotes:

☐ Spilling liquid inside the remote, shorting out the switches and circuit board.
☐ Letting batteries leak inside the remote (this usually happens only when the remote is not used for some time).
☐ Sitting or stepping on the remote.
☐ Missing battery compartment cover.

Refer to other sections in this chapter for help on dealing with spilled liquids and leaked batteries. Refer to chapter 5 for a rundown of how to open a remote control and clean it.

If spilling liquids into things like VCR remote controls is a problem in your house, one remedy is to cover the remote with clear plas-tic. The heavy-duty plastic used for sandwich and food storage bags works well. Cut the plastic to size, and wrap it around so the entire remote control is covered. Apply transparent tape to seal the plastic. You will probably want to replace the batteries before you cover the remote with plastic, so you can get a few months of use out of the protected remote before you have to uncover it to replace the batteries.

Little can usually be done to repair a remote control that's been physically abused by sitting or stepping on it. If the damage is minor, and only affects the exterior of the remote (the remote control still operates your VCR) it might be possible to glue the broken pieces back together with adhesive or plastic modeler's cement. If the circuit board inside the remote control has been cracked, you can sometimes repair it by using a glue especially designed for rebonding the insulation used in circuit boards. This glue is available at larger electronics supply houses.

Do note that repair will be difficult or impossible if the circuit board has snapped in two. A broken board requires repairing the conductive traces on the circuit board, which is time consuming and requires considerable skill. Realistically, you will need to purchase a replacement remote control if the circuit board becomes severely damaged.

A missing battery compartment cover—not uncommon in a house full of little kids—is more of a nuisance than a real problem. As long as the batteries remain firmly inside, the remote control will continue to operate. If the batteries keep popping out, apply a layer or two of black electrical tape over the battery compartment.

## Damaged tapes

A damaged VCR is one thing; damaged tapes pose a problem of their own. If a videotape is in the VCR when a calamity occurs— the deck is damaged by fire or water, for example—most often the tape is damaged as well. After any damage to a VCR you should always carefully inspect the tape to be sure it wasn't always damaged. Using a damaged tape in an otherwise healthy VCR can cause considerable problems, and it's something you will want to avoid.

Perhaps a more common damage to tapes is when a tape becomes jammed inside the VCR. With the fragile nature of the tape, most any jamming renders the tape useless and beyond repair. Invariably the jamming is caused by a dirty or worn VCR transport mechanism, and is a clarion call that your deck needs immediate attention. Although you might have every intention of servicing the VCR to keep it from wrecking other tapes, what do you do with the tape that's already ruined?

What you can do depends on the extent of the damage. The first thing is to get the tape out of the VCR. If the tape can't be extracted by gently pulling it out of the deck, **STOP IMMEDIATELY**. You will need to unplug the VCR, and remove the top cover. With the top cover off, you can more easily get your fingers inside to extricate the tape from the dastardly coils of the VCR.

Note: If you want to keep the tape, you should wear clean white cotton gloves. Gloves prevent getting skin oil on the surface of the tape, which can severely degrade it. Once the tape has been removed from the VCR, inspect it for damage, and wind it back into the tape shell (see chapter 5 for more details on working with video-tape and tape cassette shells). If the tape has been snapped in two or is severely crinkled, you will probably need to cut out the bad portion and splice the tape back together again.

Although many videotapes are minimally watchable after having been damaged inside a VCR, it's important to remember that viewing a damaged tape can be hazardous to your VCR. As explained in chapter 5, splices at any point other than the very beginning of the tape (between the clear leader and the magnetic tape) can cause damage to the spinning video heads. Damage to the heads also can result if you play a tape that has been severely crinkled. Unless you are an experienced video repair technician, replacing video heads is not something you can do, and it costs several hundred dollars to do so. There are few tapes worth the cost of replacement heads—or worse, a new VCR!

If you absolutely must save the tape—it's a record of your son's first walking step, for example—repair the tape as best you can. Then, play it once to make a copy of it on another VCR. Any side effects of the damage, like snow and static, also will be recorded, but at least you won't loose the tape completely. Of course, do remember that damage to your VCR can occur when playing a previously jammed tape just once, so proceed at your own risk.

# Troubleshooting and repairing non-VCR problems

**7**

IF YOU SUSPECT THAT ANY PART OF YOUR VCR ISN'T WORK-ing properly, take a few moments and follow the simple steps out-lined in this chapter before tearing the deck apart. Often, the problem's not a malfunction at all, but a maladjusted TV set, a loose or broken wire, or a bad videotape.

## Start at the TV

Troubleshooting the VCR begins with the TV. Consider the tuner, settings, picture controls, and reception problems.

### Proper channel and fine-tuning

If you are not getting a picture from your VCR, check the controls on the TV first. One of the most common problems is that the TV is not tuned to the proper output channel. Almost all VCRs output to either channel 3 or 4.

If your TV has a fine-tuning control, adjust the tuning knob on the set as you play back a tape on the VCR. The channel frequency generated by your video equipment could be somewhat different than what is normally broadcast by a TV or cable station, so you might have to make fine-tuning adjustments to get a crystal-clear picture.

### Proper switch selector

If your TV is connected to a video selector switch, be sure that it is dialed to VCR and not to the antenna, cable, or cable decoder box.

## Vertical hold control

If the picture on the TV rolls when a prerecorded tape is being played, check the vertical hold control on your set. If your TV lacks a vertical hold control or it isn't accessible, you might need to purchase a video stabilizer to combat the effect of the copy protection recorded on the tape (see chapter 3, "The VCR environment," for more details on video stabilizers).

## Brightness, contrast, and other controls

If the picture is dark or washed out, try adjusting the brightness and contrast controls. Colors that seem "off" could be the result of maladjusted color and tint controls on the TV. If your set has auto-color circuitry, be sure it is switched on.

## Isolating reception problems

To isolate a reception problem on your TV, hook up your cable or antenna directly to the set, bypassing everything else in your system. Take off the connection from an RF modulator, VCR, or computer, and even remove the direct audio and video connections if these have been made. If reception improves, the problem lies in an accessory coupled to the TV or the cabling that connects everything together. Reconnect each piece of equipment and its cabling, in turn, to see which one is causing the problem.

## Flickering picture

If you see a flickering picture when viewing a prerecorded tape, you might be witnessing Macrovision anticopying signals interfering with your TV. If you have an AGC (automatic gain control) switch on the TV, switch it off. If that doesn't help, try the fine-tuning knob.

The flickering picture also might occur if you have two VCRs connected to one another, even if you aren't actually recording. Depending on how you have your system connected, the Macrovision signals could be interfering with the AGC circuits in the second VCR. Try disconnecting the VCRs and viewing the tape directly from one VCR to the TV, as shown in Fig. 7-1 A and B on page 199.

*Troubleshooting and repairing non-VCR problems*

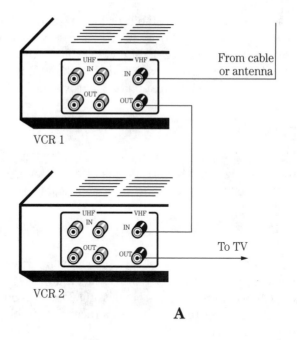

**A**

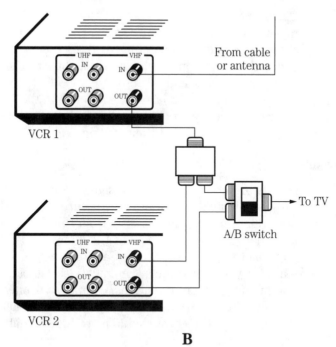

**B**

■ **7-1** *Improper connection of VCR when watching Macrovision-encoded tapes. Proper connection using an A/B switch.*

If all else fails, a video stabilizer built to counter the Macrovision signal encoding might be required. Consult your local video dealer. Sellers of video stabilizers often advertise in the video magazines. Flip through a current issue and note the ads. See the end of this chapter for more information on Macrovision.

## Program reception

If the picture you record on your VCR isn't what you think it ought to be, it might not be dirty heads or a bad tape. The program reception, either through the antenna or cable system, could be at fault.

### General

To find the cause of persistent ghosts or interference effectively, hook up your TV directly to the incoming signal source—either the cable or antenna—thus bypassing your VCR and other components in your system. If it's truly a reception problem, the interference will remain.

Check the cable(s) that connect to your TV. If you are viewing tapes from your antenna, be sure the antenna is still in good shape and that it is pointing in the proper direction. Double check the condition of the cables leading to the television. Inspect the cabling and all in-line components (matching transformers, splitters, etc.) from the antenna to the TV set. Finally, if you're hooked up to cable, call your local cable company and have them troubleshoot their system.

### Ghosts

*Ghosts* are created by the TV signal reflecting off water, water towers, tall buildings, or other structures. Your antenna receives both the original signal directly from the TV station and the reflected signal(s) that produces the ghost. Ghosts can come and go in bad weather, so wait out a storm before you do anything harsh. If you find you can't eliminate the ghost, try aiming the antenna in different directions, even away from the TV station, as shown in Fig. 7-2. The main signal will be reduced and so will the ghost, but the ghost should be reduced much more than the original signal.

Of course, if the signals from the other channels are not coming from the same source, then the reception on other channels might

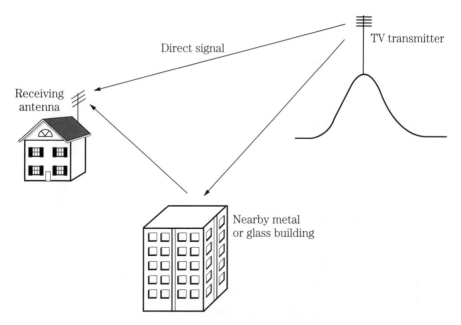

Direct signal

TV transmitter

Receiving
antenna

Nearby metal
or glass building

■ **7-2** *Ghosts are usually caused by receiving most of the direct and some of the reflected signal from the TV transmitter. The ghosts can be reduced or eliminated by carefully positioning the antenna or using an antenna that is highly directional.*

be affected. You also can purchase a *ghost eliminator*, a small device that fits between your antenna and TV. However, a ghost eliminator might not work well if the ghost signal is exceptionally strong. A more directional antenna might then be the best solution.

## Interference

Your TV is infested with interference. You've tried everything to purge the irritating wavy lines, the garbled sound, the pulsating sparkles—but nothing seems to work. Instead of correcting the problem, calls to the TV repair shop and cable company have done nothing more than set you back a week's pay. Now it looks as if the ghosts and gremlins attacking your TV are getting worse. Television interference can be an ugly thing.

Yet, chances are you don't have to live with it. If you take your time and analyze the situation carefully, most cases of television interference can be reduced or completely eliminated. Interference can be a persistent nightmare because it has many causes and symptoms. But by locating the source of your tormenting interference first, you can more easily—and therefore more cheaply—take steps to squash it.

For a quick reference on interference problems and the various causes and solutions, see the section "Interference problem/solution chart," in this chapter.

## ABCs of interference

Television interference comes in many forms and can percolate inside your TV, in other appliances in your house, from cars, within the cable system that serves your neighborhood, even from nearby broadcast stations. The sources of TV interference can be broken down into three main groups:

☐ Interference from nonbroadcast sources, like motors and faulty fluorescent lights.
☐ Interference from broadcast sources, like *CB* (citizen's-band) radios, AM radio stations, and other TV stations.
☐ Interference from within the television set.

## Interference from nonbroadcast sources

Your television is just one of several appliances you have in your house, and many of them are potential sources of interference. The most common is interference caused by ac and dc motors. This type of interference is not caused by TV or radio broadcast sources and is said to be *broadband*, meaning that the interference can span a wide range in the radio frequency spectrum. Vacuum the rug or shave with an electric shaver, and odds are that interference, in the form of tiny sparkles and horizontal streaks, pop onto the screen. Usually, but not always, the visual interference is accompanied by a buzzing, whining noise.

Generally, a well-made, properly functioning motor won't cause much interference. But in time, through wear and tear, brushes (in a dc motor) or commutators (in an ac motor) can become loose and arc. The result is electrical noise. That noise is transmitted to the TV through the ac wiring in your house. Your best bet in stopping the interference is to have the motor serviced or replaced. If that's not possible, you can help stem the interference by installing an ac line filter at the television. The filter plugs between the TV and the wall socket.

The ac line filter also helps reduce or eliminate static caused by old or faulty light-dimmer switches and fluorescent lamp fixtures. A bad dimmer switch or old fluorescent bulb (or bad ballast transformer) can cause considerable noise on the power line in your house. That noise is picked up by your TV and shows up as sparkles on the screen.

Yet another cause of nonbroadcast interference is automobile ignition. If your car has faulty ignition wiring or doesn't use resistor-type spark plugs, electrical noise can radiate from under the hood. That noise is transmitted to your TV through the air. You see the interference as random sparkles and hear it as an annoying whine that varies with the speed of the car engine. The solution is to repair the ignition wiring or install resistor-type spark plugs. If it's not your car and you can't interest your neighbor in repairing it, you can try upgrading the antenna wire leading to your TV.

Although somewhat rare, additional sources of nonbroadcast interference that can plague your TV are caused by nearby microwave ovens and from mercury-vapor security lamps. Interference from a microwave oven appears as a strange moiré or elongated $S$ pattern, generally staying at the bottom of the screen; interference from a mercury-vapor lamp shows up as constant, wavy diagonal lines. There's not much you can do to eliminate the interference from a microwave oven other than to move the TV and antenna leads away from the appliance. An ac line filter can be used to reduce or eliminate interference from a mercury-vapor lamp.

## Interference from broadcast sources

Television interference produced by such nonbroadcast sources as electric razors and faulty dimmer switches are fairly easy to track down and correct. Harder to diagnose and treat are those forms of interference caused by other broadcast sources, namely CB and amateur-radio transmitters, AM and FM radio stations, and television stations. This type of interference is said to be narrowband, meaning that the interference spans a fairly narrow swatch in the radio frequency spectrum.

Interference by citizen's-band radio takes two main forms: transmitter harmonic distortion and RF tuner overload. With transmitter harmonic distortion, your TV picture is filled with thin, slightly angled lines and garbled sound on channels 2, 5, 6, 9, and 10. The bad thing about this type of interference is that it can't be suppressed at your TV; you need the cooperation of the operator of the CB. If the CB set is in your house, attach a lowpass filter, available at most electric and radio stores, to the transmitter. If the CB is somewhere in your neighborhood, you'll have to locate it and ask its owner to attach the lowpass filter.

RF tuner overload is caused by a strong CB signal invading your TV set. The affect is frightening: the picture might black out, the sound might be completely garbled, and the entire screen might be filled with wavy interference lines. This type of interference, which can occur on most any channel, can be corrected at your TV: make sure your antenna system is in good working order, then add a CB highpass filter or fixed attenuator to the antenna terminals. The attenuator cuts down the strength of the signals reaching your TV, so you need good reception to start with.

CB transmitters outfitted with illegal amplifiers can overload your TV, causing gross distortion on the screen. Often, you can distinctly hear the CB operator through the speakers of your set. In some instances, the TV doesn't even need to be turned on; the signal radiated from the CB antenna is so strong the wiring in your TV is intercepting it.

Government regulations forbid high-powered CB transmissions. If you can locate the guilty party, provide a name and address to the local FCC (Federal Communications Commission) office. If the FCC takes action, the CB operator could have his or her radio and other gear confiscated and face considerable fines.

Amateur radio sets are another source of television interference. The symptoms are similar to that for CB transmitter harmonic radiation except that the interference can appear on any channel (which channels depends on the frequency of the radio transmitter). A highpass filter designed to block amateur radio interference can be added between your TV and antenna.

A somewhat common form of interference is caused by strong signals from a nearby FM radio station. It's most often seen on channels 5 and 6 but can appear on most any VHF channel. The visual affect is an a series of wavy, grainy lines on the TV screen. The waves change with the beat of the music on the radio. Here are three possible solutions:

☐ Install a separate antenna for receiving FM radio broadcasts. Connect it directly to your hi-fi.
☐ Separate the UHF and VHF leads right at the antenna instead of inside the house. Be sure to use a VHF/UHF signal splitter rated for outdoor use.
☐ Install an FM trap between your TV and the VHF antenna lead. If the trap has an adjustment knob, turn it until the interference is reduced or goes away.

Other television stations can cause the most disruptive form of interference. If you're midway between two stations that broadcast on the same channel, both signals might overlap one another, causing *co-channel interference*. In its mild form, co-channel interference looks like a series of dark horizontal lines—a sort of Venetian blind effect. In its worst form, the interference displays a strobing, sweeping pattern, much like windshield wipers.

Co-channel interference usually comes and goes with changes in the weather. The interference increases in warm weather when there's a change in the ionosphere; TV signals from even hundreds of miles away bounce off this high-altitude layer of the earth's atmosphere and right into your antenna. A better antenna, aimed precisely at the TV station you do want, is the best medicine against co-channel interference.

*Cross-modulation* also is caused by two different TV channels interfering with one another. The interfering picture is overlayed on top of the picture you want. Because the two pictures aren't synchronized, you often see a moving cross or bars on the screen, with an irritating herringbone effect. As with co-channel interference, a better antenna can help solve the problem. If you're close to the broadcast towers, you also might need to install an attenuator to cut down the signal strength of the interfering channel.

## TV set interference

Some forms of interference are created inside your TV set. These are generally caused by a failed component or by improper service. Depending on the problem, you'll need to have the TV set repaired to eliminate the interference.

Common forms of TV-induced interference are color oscillator interference, channel 8 *tweet*, and horizontal interference.

☐ Color oscillator interference appears as a series of steady, nearly vertical lines. It's caused by a poor connection in the color circuits of the TV. It also can be caused by an improperly installed television antenna lead-in.

☐ Channel 8 tweet appears as a series of swirling S's on channel 8 and is caused by improper antenna connections and bad wiring or components inside your TV.

☐ Horizontal interference consists of one of more vertical lines and is caused by failing components in the set. The position of the line (left, center, or right) in the picture generally indicates the faulty component.

## Cable TV interference

Just because you're on cable doesn't mean you won't be plagued by interference. In fact, depending on the quality of the equipment used in the cable system, you might experience interference on some or all of the channels. Interference is more likely if the cable system is more than 10 or 15 years old and is not maintained on a regular basis.

Generally, interference generated within the cable TV system cannot be corrected at your TV. If you suspect that the cable system is at fault, contact the cable company by phone or letter. Describe the interference as best you can:

☐ What does it look like?
☐ What channels does it appear on?
☐ Does the interference also affect the sound?
☐ Does the interference change when you adjust the fine-tuning, contrast, and color controls on your TV?
☐ Does the interference appear at all times during the day, or only at certain times?

Unless the cable company sends a repair technician to your home, you should not be charged for reporting the interference, even if the problem turns out to be caused by your equipment. Note that some cable systems are continually plagued by interference and could have little recourse in correcting it. For example, some so-called midband cable channels fall close to the broadcast frequencies used by airports. If you live near an airport, or the cable system runs close to airport transmitter towers, you could experience considerable interference. Similarly, the strong pulse from military or commercial radar could leak into the cable system, causing a buzz every few seconds.

How do you know if it's the cable system causing the interference and not your TV or cable hookup? The best way to tell is to bypass all the extra components in your video system and connect the incoming cable directly to the antenna jacks on your TV. Dial your TV as if you're using an antenna to tune into different channels. On most (but not all) cable systems, you can view the VHF channels directly—channels 2 through 13. Other channels—like HBO or Showtime—might appear scrambled or require the use of a cable-ready TV set.

## Trap, filter, and attenuator interference

Many forms of externally caused interference (that is, interference created outside the TV set) can be minimized or eliminated with a trap, filter, or attenuator. These items are available at most electronics stores, including Radio Shack, as well as many amateur and CB radio shops. If you can't find what you need over the counter, you can often special order a trap, filter, or attenuator at a TV sales or service shop. Costs range from $1.95 for a simple filter to more than $60 for a tunable trap. Be sure you know what you're getting before you buy.

### Traps

Radio signals—whether they emanate from a CB or amateur radio rig, or from an FM, AM, or TV station—are reduced or eliminated with the use of *traps*. Many traps, like an FM trap to reduce interference caused by strong FM radio signals, are tuned to just one specific frequency or sometimes a selected range of frequencies. Other traps can be tuned: after installing one on your TV, you rotate a knob until the objectionable interference is reduced or vanquished.

You can make your own fixed and tuned traps using twin lead or coaxial antenna cable. You must cut the cable to the exact length. The exact procedure and formulas for trap construction are beyond the scope of this book. A better (and safer) bet is to buy the interference traps from a local electronics dealer.

Traps are installed at the antenna terminals of your TV set with the antenna cable. For specific installation procedures, follow the instructions included with the trap.

### Filters

Filters are similar to traps except the latter are generally used to mask random electrical noise such as that caused by a faulty fluorescent light or the motor of an electric shaver. Not all filters are the same, and they're used in different ways. For example, a filter that reduces static caused by automobile ignition is installed at the antenna terminals of the TV. A filter that reduces static generated by other electrical appliances in your house is installed at the ac plug of your TV. Be sure to purchase the proper filter for your needs.

Although you can build TV interference filters, the components must be *UL* (Underwriters Laboratory) approved and carry specific electrical specifications. Unless you are experienced with RF circuit design, you're best off buying what you need at a local electronics store.

### Attenuators

Some signals can't be trapped or filtered. You use an attenuator to reduce the strength of all the TV channels if the incoming signal is overloading your TV (overload is characterized by a contrast or negative image picture, plus excessive buzzing in the sound). An all-band attenuator reduces the strength of all channels equally.

If you're picking up two adjacent channels (like 5 and 6, for example) at the same time, you'd use a tuned or tunable attenuator. A tuned attenuator reduces the strength of just one channel, such as channel 5, leaving the others relatively untouched. A tunable attenuator lets you dial the channel you want.

The amount of attenuation can be either fixed or varied. A fixed attenuator reduces the signal strength by a set amount, generally 25 or 50 percent. A tunable attenuator reduces the signal from nothing to up to 80 or 90 percent. Tunable attenuators also can be used to reduce ghosting. If you see ghosts in the picture, dial the knob on the attenuator to reduce the unwanted images.

## Interference problem/solution chart

Refer to the following quick-reference chart when attempting to eliminate unwanted interference.

### Interference: ac, dc, and universal motor noise

Caused by:   Electrical contact in brushes, commutators, or slip rings in electrical motors
Appears as:  Picture—flashing specks or streaks
             Sound—buzzing or crackling
To correct:  Install ac line filter at television set

### Interference: Fluorescent lamp noise

Caused by:   Faulty fluorescent lamp, ballast, or fixture
Appears as:  Picture—flashing specks or streaks
             Sound—buzzing or crackling
To correct:  • Replace lamp or ballast
             • Repair faulty fluorescent lamp fixture
             • Install ac line filter at television set

### Interference: Dimmer switch noise

Caused by:   Old or faulty dimmer switch (generally when switch is set too low)
Appears as:  Picture—flashing specks or streaks
             Sound—buzzing or crackling

To correct:  • Replace dimmer switch
             • Place ac line filter at television set

### Interference: Automobile ignition

Caused by:   Faulty wiring in automobile; use of nonresistive spark plugs in automobile
Appears as:  Picture—specks and streaks
             Sound—buzzing or whining noise
To correct:  • Repair or upgrade automobile ignition system
             • Replace antenna cable with shielded coaxial type
             • Move antenna and cable away from traffic

### Interference: Microwave oven; RF heating

Caused by:   Home or industrial RF heating units, such as microwave ovens.
Appears as:  Picture—S-shaped moiré patterns in lower portion of screen; might roll vertically through picture
             Sound—none
To correct:  Install highpass filter on antenna leads of TV.

### Interference: CB transmitter harmonic distortion

Caused by:   CB radio
Appears as:  Picture—diagonal lines on channels 2, 5, 6, 9 and 10; lines might move as CB operator speaks
             Sound—garbled
To correct:  Install lowpass filter on CB radio

### Interference: CB transmitter RF tuner overload

Caused by:   CB radio
Appears as:  Picture—high-contrast or completely dark picture
             Sound—garbled
To correct:  • Reorient antenna away from source of CB signals
             • Use coaxial antenna cable
             • Install highpass CB filter at antenna terminals of TV set

### Interference: Amateur radio fundamental and harmonic radiation

Caused by:   Amateur radio
Appears as:  Picture—diagonal lines, fluctuates as radio operator speaks through microphone
             Sound—garbled; might hear radio operator
To correct:  • Install highpass filter at antenna terminals of TV set
             • Install tuned trap at antenna terminals of TV set

### Interference: FM broadcast

Caused by:   Nearby FM stations

Appears as:  Picture—grainy, fluctuating lines; fluctuating diagonal lines
Sound—none or raspy noise

To correct:  • If the hi-fi also is connected to the TV antenna (through a suitable band splitter), install separate FM antenna
• Install FM signal trap at antenna terminals of TV

### Interference: co-channel interference

Caused by:   Two signals on same channel overriding one another.

Appears as:  Picture—thick horizontal lines (Venetian blind); pulsating moving lines (windshield wiper)
Sound—none or garbled

To correct:  • Orient antenna towards direction of desired channel
• Replace antenna with more directional model

### Interference: Cross modulation

Caused by:   Two signals from adjacent channels overriding one another

Appears as:  Picture—herringbone effect; video cross-point pattern slowly moving through picture
Sound—none or slight raspiness

To correct:  • Orient antenna towards direction of desired channel
• Adjust fine-tuning
• Install tuned channel attenuator to block interfering channel (the interfering channel is the stronger one, the channel that appears over the one you want)

### Interference: Color oscillator interference

Caused by:   Improperly connected antenna system; fault in color circuits inside the TV

Appears as:  Picture—steady, slightly canted lines
Sound—none

To correct:  • Repair or replace faulty antenna components
• Repair or replace faulty color circuits in TV

### Interference: Channel 8 tweet

Caused by:   Bad antenna connections; fault inside TV

Appears as:  Picture—S-shaped herringbone pattern on channel 8
Sound—none

To correct:  • Repair or replace faulty antenna components
• Repair or replace faulty circuits in TV

**Interference: Horizontal interference**

Caused by:  Fault inside TV

Appears as:  Picture—thick vertical (sometimes bowed) black lines (usually appears only on weak channels)
Sound—none

To correct:  Repair television set

# VCR cables and controls

Like the TV, problems with a VCR are often caused by improper connections and maladjusted controls. Make sure all of the cables are attached properly and that each one is connected to the right input or output. Double check the cable connections by referring to your operator's manual, especially after disconnecting your VCR to move or clean it.

Go through the same procedure outlined previously for making sure you are on the proper channel. If you are connected to cable, the decoder or converter box might have a channel 3 or 4 output, so you must dial your VCR to that channel to receive a picture. Also check the TV/VCR switch. This switch allows selection between VCR playback and over-the-air program viewing through the VCR tuner.

# Tracking control

The tracking control on the VCR helps assure synchronization with tapes made on other machines. If the picture is snowy, particularly on the top or bottom of the screen, adjust the tracking control. In some cases, the control will have little or no effect. But just because the control doesn't work doesn't mean that the VCR is at fault. In nine times out of ten, the tape will be faulty or recorded on a deck that is poorly aligned. You can isolate the problem to the VCR by trying other tapes in your collection.

If your VHS VCR is equipped with hi-fi audio, you might need to adjust the tracking to "tune in" the sound track. This tuning is especially needed if you are viewing a tape recorded on another machine, such as a prerecorded movie. If the tracking is misadjusted, the VCR will alternate between the hi-fi track and the linear audio track, causing annoying popping and a change in the audio level.

If your VCR is equipped with VU meters for the sound level, use them to adjust for proper tracking. Note that in some cases, you won't be able to adjust the tracking for both video and hi-fi audio at the same time; rather, find a midpoint where both are acceptable.

Many of the newer VCRs have automatic, or digital, tracking. The tracking is adjusted by the VCR automatically by detecting the strongest video signal possible. The telltale noise of poor tracking reduces the strength of the signal; circuits in the VCR automatically adjust the tracking until the signal is the strongest.

For automatic tracking to work, the tape must be in fairly good shape to begin with. A margin tape might not track properly, no matter how much the tracking is adjusted. On these, a VCR with automatic tracking might seek out the optimal tracking and never find it. The result is that the tracking can get better and worse as the VCR attempts to compensate.

If your VCR cannot adequately track a tape automatically, you might need to override the automatic tracking feature and adjust the tracking manually. Once you override the automatic tracking, the VCR will be prevented from further adjusting the tracking by itself, as long as the tape is played. Most VCRs with automatic tracking lack a separate tracking control. Instead, the manual tracking override buttons are shared with other buttons on the VCR, most often the channel UP/DOWN buttons (check the manual that came with your VCR for more details on how to manually adjust the tracking). When playing a tape, push the tracking buttons until the picture is the clearest.

## TV/Video switch

VCRs have a TV/Video switch that lets you manually switch between over-the-air viewing through the VCR and tape viewing. When in TV position, you see the channels on your TV as they are received by the tuner in the VCR. Your set is tuned to the output channel of the VCR, usually channel 3 or 4. In Video position, you see the playback of the tape.

Depending on how you have your VCR connected to your TV and video system, pressing the TV/VIDEO switch when watching a tape might switch you to over-the-air viewing. In older models, the position of this switch is not reset each time you press the PLAY button of your VCR. Although your VCR is working properly, your TV will continue to receive the broadcast channel through the VCR's tuner, not the program on the tape. The moral is always to check the position of the TV/Video switch. When you are not getting a picture, push it to see if the taped program comes on.

Because they are all electronic, on most of the newer models of VCRs the TV/VIDEO switch is reset when the power to the VCR is turned off (the switch reverts to the TV position). When you turn the VCR back on, the switch stays in the TV position. You must manually change the switch to the VIDEO position.

## Remote interference

Assume you are trying to play a tape, but pushing the buttons on the VCR has no effect. Bad tape? Maybe. Bad VCR? Probably not. What else could cause the problem? The remote control. If your VCR has a remote control, either wired or wireless, its controls usually override the main controls on the front panel of the deck. Locate the control and make sure that no books, plates, magazines, or other debris are pushing on the buttons, improperly commanding the VCR.

Sometimes, remote buttons get stuck, so even though the controller seems healthy on the outside, it is still interfering with the operation of the VCR. If the remote is the wired type, disconnect it from the deck. If it is the wireless type, take it into another room, remove its batteries, or cover up the front of it to block the passage of infrared light.

Three types of remote controls are used on VCRs: wired, infrared, and UHF.

☐ Wired remotes are passé, and are only found on older models. The remote is tethered to the VCR by a wire (this wire also can break inside, causing the remote to stop working).

☐ Infrared remotes, by far the most common, use invisible infrared light to control the VCR. The remote control needs to be in close proximity to the VCR to work properly, so simply taking the remote to another room can determine if the remote is causing the interference.

☐ UHF remotes are available as an extra-cost option on some VCRs and as replacement remotes. They work by sending a radio signal (in the UHF region of the radio frequency spectrum) to the VCR. The receiver is either built into the VCR, or is contained in a separate unit that you attach to the VCR. Because UHF remotes are designed to work between rooms (including through walls), you must remove the batteries to the remote to determine if the remote is causing the interference.

213

## Using a known good tape

If your VCR is suffering from one or more picture and sound problems when playing tapes, test the deck with a known good tape. (A known good tape is one you've used recently that seemed okay or a relatively new store-bought prerecorded tape. To be sure it is in good condition, try it using a different tape player). Always play a known good tape before passing judgment on your VCR.

There are two general forms of out-of-order cassettes: broken cassette shells and damaged tape.

### Broken cassette shells

A broken cassette shell is one that is cracked or internally damaged so the tape no longer moves through it smoothly. A broken roller might prevent the tape from coming out, and this will cause the tape to jam inside the cassette. Similar problems occur when the cassette hatch (the door on the front of the shell) snags with the tape or when the internal reel locks inside the shell fail to disengage (for a complete description of the inside of VHS, VHS-C, Beta, and 8mm cassettes, see chapter 2). You might be able to repair the cassette by moving the tape to another shell, as discussed in chapter 5.

### Damaged tapes

A damaged tape is rarely repairable. *Damaged* means a tape that has been folded, spindled, or mutilated in such a way that the VCR can no longer read the signals impressed on it. A stretched tape, as shown in Fig. 7-3, is not only dimensionally thinner than a good tape, but it buckles and prevents the video, audio, and control-track heads from reading the information.

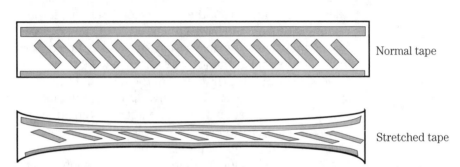

Normal tape

Stretched tape

■ **7-3** *Stretched tape deforms the signal area and impairs playback.*

Tape edges that have been cut or scored can cause audio or tracking problems. In all formats, the linear audio tracks are along the top of the tape and the control track is along the bottom.

A problem with the tape shell or VCR can scratch the tape. If the problem is serious enough, the scratch can remain for the entire length of the tape (much like a scratch in film caused by a dirty projector). The scratch shows up as a streak of white or black on the screen and can cause the video heads to become prematurely clogged.

A tape that has spilled out of the cassette, either inside or outside the VCR, is subject to mutilation. At the very least, the tape will be folded and creased along its length. You should not play a tape damaged in this manner because it can actually snap off a video head. If the creasing is light, you can use the tape, but it's better to make a copy and throw the original away.

Another alternative is splicing out the bad portion. But unless the splice is very well made (as discussed in chapter 5), you run the risk of damage to the video heads. The reason is that the video heads don't just skim across the tape, they actually protrude and penetrate into the tape, causing it to bulge out. A gap in the splice is like a pothole for a car—the head can dig into the gap and break off. Tape jamming and spillage can occur if the tape ends are incorrectly aligned, as shown in Fig. 7-4.

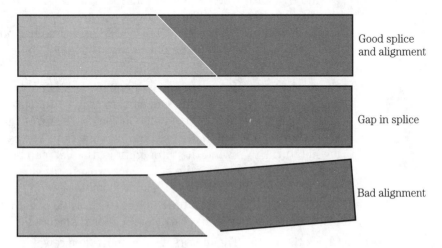

Good splice and alignment

Gap in splice

Bad alignment

■ **7-4** *Avoid splices that are improperly aligned or leave gaps between the tape ends.*

# Bad tape rewinder

Your VCR is fine; it's the tape rewinder that's sick. A tape rewinder saves wear and tear on your VCR by rewinding the long lengths of tape back to the beginning. To use the rewinder, you just insert the tape, push down the access door, and the tape rewinds. The access door pops open when the tape is finished rewinding.

Most tape rewinders are rather simple devices, and they aren't made with lasting quality in mind. Depending on how much you use it, most rewinders will last a few years before they will have to be replaced. The rewinding motor might burn out or the belts inside might wear out. Because the average tape rewinder costs about $10 to $15, it's often cheaper simply to toss it and buy a new one.

However, you might not wish to junk your tape rewinder so quickly, particularly if it's a fancier model that costs more. If your tape rewinder is sick you might be able to correct it.

Here are things to check if your tape rewinder isn't working properly.

*Caution: Be sure to unplug the tape rewinder before attempting to service it!*

1. Check that the rewinder is plugged in. Many rewinders use an external power pack; one end of the pack plugs into the wall, the other end plugs into the rewinder. Make sure both plugs are securely in place.

2. Check the actuator switch. The actuator switch turns on the rewinding motor when you insert a tape. The switch is often the "leaf" or lever type. The lever can become bent, requiring you to straighten it out. Also, check that the switch has not become broken, and that the power leads have not come loose.

3. Check the belts. Many tape rewinders use a system of belts and pulleys in the rewinding mechanism. The motor turns a pulley, which is attached to a belt. On the other end of the belt is another pulley, which drives the rewinding hub. The rewinding hub turns to rewind the tape. In time, the belt can become broken, worn, or dirty. If the belt is not broken, try cleaning it with belt cleaner (the type of cleaner for phonograph players is sufficient). If the belt is broken, measure it and order a replacement. See appendix A for sources for replacement belts.

4. Check the motor. The motors used in most tape rewinders is under-powered for the job. This means they don't put out much

of a magnetic field that might accidentally erase the tape, but they also tend to burn out after some use. If the motor is burned out, it's time to buy a new rewinder. A replacement motor would cost more than the price of a new rewinder!

## Macrovision mayhem

*Macrovision* is a system designed to prevent casual VCR-to-VCR copying of prerecorded tapes in the home and is used in more than 90 percent of all prerecorded video titles. More than 200 million videocassettes have been recorded with the Macrovision anticopy signal, and that number increases every day.

Although no one can blame producers and distributors of video programming for wanting to protect their copyrights against illegal duplication, since its inception the Macrovision process has been sharply criticized for impairing the quality of tapes. Viewers—and even some video dealers—have complained of substandard sound and picture on tapes recorded with Macrovision. So say the critics, these goblins of vexed video manifest themselves as a blinking, pulsating, or rolling picture, grainy video, and even strange ghostly glows that wander about the bottom of the TV screen.

Macrovision (the company) and the movie studios have downplayed any destructive tendencies of the anticopying process. Any problems, they say, are caused by faulty wiring in the viewer's home, by a maladjusted TV set, or by errors in tape duplicating, entirely apart from the Macrovision encoding process.

The proliferation of Macrovision and the alleged side effects make this a subject to cover in a book about VCR maintenance and repair. If your VCR is having trouble playing a prerecorded movie, it might not be the deck's fault, but Macrovision. But how do you really know if Macrovision is to blame? How do you determine if the tape is flawed or your VCR needs adjustment? Here's how Macrovision works, how it can affect your TV, and how to solve Macrovision-related problems.

### What is Macrovision?

The basic idea behind Macrovision is to render videotapes uncopyable from one VCR to another, yet allow you to watch tapes unimpaired on your TV set.

Macrovision isn't the first anticopying process to find a home on consumer videos. One system, called Copyguard, was originally applied to prerecorded videotapes in the late 1970s and early 1980s. But Copyguard more often than not bedeviled TV sets. The system was finally abandoned: too many viewers complained of jittery pictures and an irritating bending at the top of the picture.

The concept of the Macrovision anticopy process is relatively simple. Electrical pulses of specific strength and duration are added during selected portions of the video signal. The Macrovision pulses are placed in the vertical blanking interval, a segment of the video signal that controls the vertical synchronization of the TV set.

The first part of the vertical blanking interval is dedicated to the vertical sync pulses, those signals that actually control the synchronization of the TV set. The second part of the vertical blanking interval consists of 12 extra picture lines (recall that a TV picture is really composed of many lines stacked upon one another on the screen). These lines comprise the dark bar you see when the vertical hold of your TV is misadjusted. Some of these extra lines in the vertical blanking interval are used for special applications, like close captioning and color reference.

The Macrovision pulses take the place of six or more of the unused lines in the vertical blanking interval. Tapes made with an early version of Macrovision used pulses of constant strength, but later refinements to the Macrovision process introduced pulses of varying intensity.

These pulses—whether constant or varying—are designed to upset the automatic gain control (AGC) circuits in a recording VCR. The AGC automatically tracks the level of the video signal, and makes adjustments so the picture is the proper brightness.

With the Macrovision signals added, the AGC circuit is fooled into thinking the picture is really lighter or darker than it really is. The typical result, as viewed on a protected tape copied from another VCR, is an overly dark picture, or more commonly, a picture that alternates between light and dark scenes.

## What Macrovision isn't

Macrovision isn't a terribly complex process, but it does involve technologies unfamiliar to most video enthusiasts. Because of technical ignorance, many consumers erroneously believe Macrovision is many things that it is not. Macrovision does not:

☐ Destroy the tape after a certain number of plays or after a certain time

☐ Alter the sound portion of the tape
☐ Fill the screen with snow instead of a watchable picture
☐ Harm your VCR or TV

To be fair, consumers are not all to blame. Some videotape companies have printed idiotic warnings of impending doom on their tapes. The warnings imply or even clearly state that attempting to copy the original tape will result in erasing or destroying it. These cautions are on tapes recorded with and without the Macrovision anticopy process.

## When things go wrong

The developers of Macrovision estimate that upwards of 85% of all VCRs cannot effectively record a protected tape. The author's informal survey has shown that percentage to be somewhat lower, on the order of 60 to 70 percent. Depending on your VCR and TV, you might never even see the effects of Macrovision, even if you copy prerecorded tapes encoded with it.

By the same token, it is not unreasonable to assume that if Macrovision does not affect some VCRs, it could have a scurrilous influence on some TV sets. Macrovision claims this is not the case, but many viewers insist otherwise. Who's right?

Both are. First a little history. The Macrovision process was first applied to the movie "The Cotton Club" in mid-1985 and has evolved since. Since then, there have been no fewer than six different formats of Macrovision, the first five taking place between May 1985 and November 1986. Since late 1986, the Macrovision process has remained stable, with only minor apparent modifications. It is this last format that is most compatible with the nation's TVs and is largely responsible for the general decline in complaints against Macrovision-encoded tapes.

The first Macrovision process, as used on "The Cotton Club" (it is not known to have been used on others), consisted of a series of constant-strength pulses placed immediately after the equalizing pulses during the vertical blanking interval. The Macrovision pulses were very strong—too strong perhaps, and it caused bright white lines in many television sets.

Two other forms of a slightly modified version of the original Macrovision format were used—with "Torchlight," "The Sure Thing," "Mask," "Invasion:USA," and others—until about mid 1986. These two subformats were identical to the original, except the Macrovision pulses were placed on later lines during the vertical blanking interval and were not as strong. If you rent or buy any of these tapes and are plagued with the problems mentioned, the culprit could very well be Macrovision.

Some of these earlier formats injected a Macrovision pulse on line 19, which is "officially" reserved for the *VIR* (vertical interval reference) signal used by some TV sets and broadcast on most television channels. On sets that use this signal as a reference, a Macrovision pulse there caused serious shifts in color. For obvious reasons, Macrovision pulses are no longer placed on line 19.

A fourth format, used in "The Blues Brothers," among others, added extra pulses before the vertical blanking interval—in effect, at the very bottom of the screen. Although this must have seemed like a good idea to the purveyors of Macrovision, these innocuous pulses proved malevolent when viewing tapes on certain models of Sony TVs (including the KV-25XBR and KV-2665), as well as sets using some Sony-built picture tubes. The Macrovision pulses placed before the vertical blanking interval caused an odd, crescent-shaped glow in the lower-right portion of the TV. The glow is caused by an optical reflection of the Macrovision signal back into the picture tube.

Apparently, the first versions of Macrovision were not as effective with all VCRs as the movie studios would have liked, and Macrovision made a drastic change to their system in time for the legendary release of "Back to the Future." This movie was the first major release to use the super-Macrovision system, where some of the pulses change in strength over time, and others remain constant. Rather than keep all the pulses at the same high level, which produces an overly dark but constant picture when an illegal copy is attempted, the varying pulses cause a more annoying blinking or light-to-dark strobing effect.

"Back to the Future" also retained the Macrovision pulses immediately before the vertical blanking interval, and the telltale crescent glow remained to haunt thousands of Michael J. Fox fans. The process, as used on "Back to the Future," as well as dozens of other titles released around the same time, also occasionally exhibited a rainbow at the top center area of the picture (generally visible only during dark scenes) and a jumpy, jittery picture (during light scenes). Macrovision now readily admits these shortcomings of their older processes.

Macrovision "fine-tuned" the process on November 1, 1986 to eliminate the glow, rainbow pattern, and shaky picture. The "fine-tuning" consisted mainly of removing the Macrovision pulses that appear before the vertical blanking interval.

## TV problems

Macrovision is designed to discourage VCR-to-VCR copying but it is engineered to display a perfect picture on your TV screen. Not all TVs are created equal, however, and certain television sets—particularly the older models made until the mid 1970s—use AGC circuitry susceptible to almost any unexpected modification in the input video signal.

Older TVs, especially the RCA XL-100 models as well as sets manufactured by the mid 1970s by Curtis-Mathes, Sylvania, Magnavox, and Zenith, exhibit an unstable picture even when the vertical blanking interval signals are modified only slightly. If you have one of these older TVs, it might be more prone to the effects of Macrovision than your neighbor with his late-model set.

How many of these older sets are still in use? That's hard to tell, but a recent survey by the Electronics Industry Association—a coalition of consumer electronics manufacturers—estimates the lifespan of the average TV is 15 years. Certainly, then, there are several million TV sets still in use that are 15 years old or more, and each one is a possible (though not necessarily likely) candidate for Macrovision mishaps.

Compatibility also depends on how you use your TV. If your TV has a sensitive automatic color control circuit that you regularly leave on, you might find it necessary to turn it off to view an acceptable picture when watching a Macrovision-encoded tape.

## The Macrovision presence

Macrovision (the company) once released a two-page circular titled "Viewing Problems Unrelated to Macrovision." This circular details various problems some viewers have encountered when watching protected tapes. Problems like dark pictures and pulsing video are not caused by Macrovision, the paper said, but by a faulty VCR, TV, or videotape.

But the title of the circular is misleading. Many of the symptoms of viewing problems are related to Macrovision, but they might not be the sole fault of the anticopy process. The mere presence of Macrovision can cause problems that might not occur if the anticopy signals were not on the tape in the first place.

Tape duplicators apply the Macrovision signals to tapes as requested by the movie studio or distributor, turning it off when it's not wanted. Placing Macrovision on a tape requires that the AGC switch be turned off on the recording slave VCRs. If the tape operator forgets to flick the switch, tapes made on the VCR might exhibit signs of Macrovision spoiling—alternating light and dark scenes, or worse. This problem partially accounts for the occasional bad apple among hundreds of good tapes.

Two VCRs connected together could cause the familiar light and dark pulsations, even if you're not copying a tape. When playing a tape that feeds to a second VCR then to the TV set, the AGC circuits in the second VCR will react to the Macrovision signals.

Should the vertical height of your TV set be improperly adjusted so any portion of the vertical interval is shown on the screen, the Macrovision pulses could intrude into the picture. The intrusion causes light and dark flashes and sometimes a rolling or jitter picture. The problem is particularly troublesome on tapes made before November 1986 that use Macrovision pulses before the vertical blanking interval.

You're watching an illegally duplicated tape originally recorded with Macrovision. Instead of a normal picture, the copy is permanently recorded with the light-dark flashes caused by Macrovision.

Early models of the TeleCaption II close-captioning device were not compatible with Macrovision. Users of these early models who viewed tapes like Disney's "Sleeping Beauty" (which has both close captioning and Macrovision) found the captioning information was not displayed on the screen. The problem, apparently created in the design of the captioning device, was corrected in later models of the TeleCaption II.

## Remedies and solutions

There are a number of steps you can take if you're having trouble with a tape and you think Macrovision is to blame. First, make sure your VCR and TV are operating properly and all connections between the two are secure. Adjust the tracking control on your VCR if the picture looks grainy or if the picture appears snowy. Clean the heads in the recorder if all your tapes exhibit snow or the picture looks weak. If possible, check your system against a known good tape. Then check the following should the problem still persist:

☐ If you have two VCRs and they are hooked in daisy-chain fashion (the output of the first VCR connects to the input of the second before finally attaching to your TV set), play the Macrovision-encoded tape on the second VCR. You also can wire an A-B switch so you can select a direct connection to the TV from either VCR.

☐ If your video system includes any enhancement assessors (like character generators, special-effects generators, or image enhancers), try bypassing them.

☐ If the vertical height of the picture is set too low and you see any portion of the vertical interval on the top of the screen or blank lines along the bottom, have your set serviced. Some models have a vertical height control on the back where you can make the adjustment yourself.

☐ If the picture is excessively bright or washed out, try turning the automatic color control off or manually adjusting the COLOR and TINT controls. These adjustments are sometimes necessary on any tape, not just those with Macrovision.

The problem still remains? If the tape is a rental, take it back and ask for a credit for another copy or a refund. Explain why you were unable to view the tape. If you purchased the tape, take it back and ask for a new copy (refunds are rare from dealers on purchased tapes). Again, explain the problem.

Although most video dealers will give you credit for viewing hassles—as long as you don't abuse their patience—not all are so understanding (though you have a right, under the consumer protection laws of most states, to receive a replacement or refund for defective merchandise). If your tape woes are not answered satisfactorily by the dealer, write the distributor or movie studio, explain your problem, and ask for a replacement.

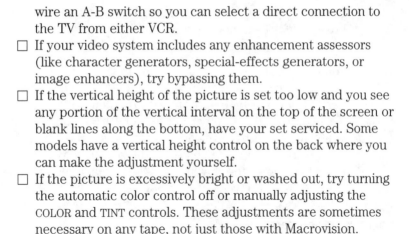

223

If the tape uses one of the older Macrovision formats and has not been re-released, it's likely that even a replacement won't help. After trying one more tape to be sure the first tape wasn't merely a lemon, you are better off asking for your money back.

## Identifying Macrovision tapes

How can you tell if a videotape has been recorded with Macrovision? Most studios and distributors refuse to indicate on the videocassette label that Macrovision has been used (although a few individual tapes, most notably "The Sure Thing," include a "Macrovision encoded" message or something similar on the box cover or the tape itself). However, if your TV has a vertical hold control, you can turn it slightly one direction or another to bring the vertical blanking interval bar into view. That will reveal the Macrovision signals, if they are there. Instead of a solid dark bar, Macrovision-encoded tapes exhibit a large checkerboard of five light and dark boxes. On all but the early Macrovision releases, portions of the boxes might flash between light and dark or slowly alternate between black, gray, and white.

You can identify a tape made prior to the November 1986 "fine-tuning" by closely examining the right side of the vertical blanking bar. On tapes made before this date (except for the very early releases during the summer and fall of 1985), you'll see a distinct white line on both the top and bottom edges of the bar. These lines extend from about the center of the screen to the right side. On tapes made after the November 1986 fix, you'll see only one line on the bottom of the vertical blanking bar.

## Macrovision "busters"

Okay. You rewired your video system so original tapes you play on your VCR are sent directly to your TV set and not first to another VCR or other piece of video gear. You've recently had both your TV and VCR serviced, and they're in top working order. Yet the deleterious effects of Macrovision still haunt you and ruin your legitimate viewing enjoyment.

You might find relief in a Macrovision buster, a "black box" that defeats the anticopying process by removing the irksome Macrovision signals. But take note: such black boxes are hard to find. Macrovision has brought or threatened patent infringement lawsuits against most companies that have attempted to market such devices. Macrovision coyly patented all the ways they could think of to defeat their system; that patent (number 4,695,901) was awarded on September 22, 1987.

Among the makers of black boxes who have withdrawn from the Macrovision-buster business are Xantech, Vidicraft, Elephant Electronics, MFJ Enterprises, Hobby Helper (who designed a black box kit for the December 1987 issue of Radio-Electronics magazine), Showline, and Digital Technologies.

Despite the threat of lawsuits, Macrovision busters are still around, but it could take some detective work to find them. Look for ads in the video and electronics magazines, or visit smaller, independently owned video dealers in your neighborhood.

Keep in mind that, like building or selling, just using a black box to defeat Macrovision could be patent infringement. Perhaps worse is that many black boxes tested degrade the picture, sometimes substantially, while removing the Macrovision pulses. And, the black box might not work satisfactorily on all protected tapes because of the change in the Macrovision process over the years.

## Troubleshooting checklist

To recap, before you tear open your VCR, check these components of your video system first:

- ☐ TV operation (including all controls)
- ☐ Anticopying signal on prerecorded tape
- ☐ VCR-to-TV cables
- ☐ Antenna or cable system hookup
- ☐ Reception
- ☐ TV/VCR switch
- ☐ Tracking control
- ☐ Remote interference
- ☐ Tape quality

# Troubleshooting techniques and procedures

**8**

THE WORD *TROUBLESHOOTING* LITERALLY MEANS AIMING AT trouble and firing away until you hit the bull's-eye. In a more practical sense, troubleshooting means locating and eliminating sources of problems but doing so in a logical and predescribed manner.

Troubleshooting is the basis of electronic and mechanical repair, and a thorough grasp of its techniques and procedures is important. Just as you can't hope to shoot at a target with a blindfold over your eyes, you cannot wildly attack a problem in your videocassette recorder and pray that you find the solution through luck. Troubleshooting lets you approach the problem from all angles and zero in on the cause with the least amount of wasted time, energy—and most of all—money.

This chapter details the concepts behind VCR troubleshooting techniques and how to apply them to actual hands-on procedures. If you are already familiar with electronic and electromechanical troubleshooting, the information in this chapter might seem old hat to you. If so, skip to the next chapter. It contains troubleshooting flowcharts that you can use to pinpoint common problems with your VCR. If you are not already familiar with basic troubleshooting techniques and procedures, be sure to read this chapter, as it contains useful information you won't want to miss.

## The essence of troubleshooting

You can better understand the role that troubleshooting plays in the repair of VCR ailments by using a more familiar concept: figuring out what's wrong with the family car. Assume you go out to your car one morning, turn the key, and the car won't start. The battery turns the engine, but even after 10 or 15 tries, the car simply won't start, and every minute you spend cranking the engine is every minute you are late for work.

You could open the hood, tear out the engine and rebuild it, or you could replace random parts, thinking that it's got to be one of them that's causing the problem. You know better of course, and you stop and think for a minute: why won't the car start? The engine turns over, so you can rule out a bad battery. But the problem could be in the ignition system, the fuel lines, or a number of other subsystems. In fact, there are several possible causes:

☐ The engine is flooded
☐ The spark plugs are fouled
☐ The engine isn't getting gas
☐ The engine timing is off
☐ The spark isn't strong enough
☐ There is no spark
☐ The engine isn't getting enough air because the choke is closed
☐ . . . and so forth.

Once you have identified the possible causes, you can take corrective steps on a one-by-one basis. You start with the most common or probable cause and work your way down. In most cases, failure to start the engine is caused by flooding. To remedy this, open the hood and remove the air filter and perhaps a couple of spark plugs. The gas evaporates and you start your car. If the problem still persists, you go to step two, and so forth.

This example outlines the basic three-step process to troubleshooting, and it applies to VCRs as it does to hard-starting cars.

1. Analyze the symptom—"car won't start but engine turns over"—and develop a list of possible causes.

2. Arrange the causes in order from the most likely to the least likely.

3. Start at the top of the list (most likely cause), and by a process of elimination, inspect, test, or otherwise rule out each possibility until you locate the problem.

Once you find the problem, you can clean or repair the faulty component.

# Troubleshooting flowcharts

It's sometimes easier to visualize the troubleshooting process by using flowcharts. These are graphic representations of the possible causes and their solutions. The exact form of the flowchart can vary, but the basic information they contain is the same.

One example of a troubleshooting flowchart is shown in Fig. 8-1. on the next page. The chart is labeled with the problem (or symptom), which in this case is "Engine turns over but won't start." Below the title is a set of boxes. The boxes are stacked vertically, and each box on the left-hand side contains a possible cause for the malfunction, usually a bad or dirty part or subsystem. The boxes are organized from the most likely to the least likely, so you won't waste your time with an unusual fault when the actual cause is really common. In each level are two or three boxes. These boxes contain the suggested remedy, and in some cases, the procedure for testing the suspected part or subsystem.

Arrows connect the boxes in such a way that by thinking of the boxes as questions, you can answer them simply with a Yes or No response and thereby navigate yourself around the chart. Each No answer to a possible cause brings you down to the next level. Answer Yes to a cause, and you move laterally as you make tests and repairs.

In actual practice, however, you don't really know where the problem originates until you test each possible cause. Therefore, at each level you initially answer with a Yes response until you have the opportunity to test it out. If the test proves negative (not the cause), you move down to the next level.

# Using the troubleshooting charts in chapter 9

The troubleshooting flowcharts in chapter 9 follow the same form and logic as the example at the beginning of this chapter. If your VCR isn't working properly, find the chart that most closely matches the symptoms of the deck. Some machines exhibit multiple symptoms, and you might need to consult several charts as you attempt to pinpoint the cause.

Once you've identified the proper chart to use, start at the top level and work your way down, eliminating those causes that you're sure are not the source of the problem. Double check your work to make sure you haven't missed something, and don't forget the obvious. By far, the majority of problems with VCRs are caused by seemingly

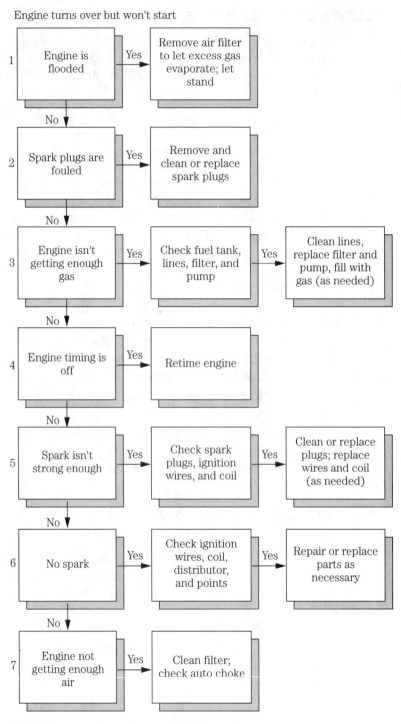

Engine turns over but won't start

| | | | | |
|---|---|---|---|---|
| 1 | Engine is flooded | Yes → | Remove air filter to let excess gas evaporate; let stand | |
| | No ↓ | | | |
| 2 | Spark plugs are fouled | Yes → | Remove and clean or replace spark plugs | |
| | No ↓ | | | |
| 3 | Engine isn't getting enough gas | Yes → | Check fuel tank, lines, filter, and pump | Yes → Clean lines, replace filter and pump, fill with gas (as needed) |
| | No ↓ | | | |
| 4 | Engine timing is off | Yes → | Retime engine | |
| | No ↓ | | | |
| 5 | Spark isn't strong enough | Yes → | Check spark plugs, ignition wires, and coil | Yes → Clean or replace plugs; replace wires and coil (as needed) |
| | No ↓ | | | |
| 6 | No spark | Yes → | Check ignition wires, coil, distributor, and points | Yes → Repair or replace parts as necessary |
| | No ↓ | | | |
| 7 | Engine not getting enough air | Yes → | Clean filter; check auto choke | |

■ **8-1** *Flowchart for engine-trouble diagnosis.*

*230*

innocent things like an old tape, a dirty head, a dirty switch, or a selector on the deck or amplifier that's not set correctly.

Whatever you do, avoid the temptation to tear your VCR apart before you adequately identify the cause of the problem. Disassembly and removal of any part, even the top cover, should only be done after you've eliminated the other causes. Remember that you cannot repair or replace some components in VCRs without special tools and techniques, so make it a point to avoid removing parts unless you absolutely have to.

## What you can and can't repair

You can try to repair anything in your VCR. It's yours, you own it, and you're free to do anything you like with it. Many problems are easily handled by the home technician, and there is no reason why you should not diagnose and fix them yourself. However, there are a number of malfunctions that are best left to a repair technician who has the service literature and special alignment and test jigs for your specific deck.

The following sections contain a general list of the components and assemblies you can repair in your VCR and a list of those you should refer to a qualified technician. This list is not absolute by any means. Your level of expertise and the amount of service material and tools you have or can get influence the types of repairs you can perform.

### What you can do

- [ ] Clean and inspect the video heads
- [ ] Clean and inspect the stationary-erase, audio, and control-track heads
- [ ] Clean the exterior and interior to remove dust, nicotine, and other contaminants
- [ ] Clean and replace belts and rubber rollers
- [ ] Inspect and replace springs
- [ ] Lubricate gears, shafts, and other moving parts to prevent them from grinding
- [ ] Replace or resolder broken wires
- [ ] Clean or replace dirty or broken switches, including front-panel switches
- [ ] Test for proper operation of the capstan, take-up reel, and loading-mechanism motors, and replace them if necessary (these usually do not require critical alignment)

☐ Replace the main printed circuit board (PCB) and ancillary boards (special handling of the boards required)
☐ Replace common electronic components (resistors, capacitors, diodes) on the PCBs, but only if the exact original value is known

### What you should not do

☐ Disassemble anything on the video-head drum
☐ Replace video heads
☐ Adjust the heads and guide rollers inside the deck
☐ Adjust trimmer potentiometers on the main PCB without the use of a schematic and an oscilloscope
☐ Replace integrated circuits, transistors, or any other component where an exact replacement or substitute cannot be guaranteed

## What to do if you can't repair it

Should you find that repairing your VCR is beyond your technical expertise and resources, don't fret. Make a note of the problem and the suspected cause, reassemble the deck, and take it to be serviced. By performing the basic troubleshooting procedures yourself ahead of time, you make the repair technician's job that much easier. You might not get a reduction in the labor charge, but if the problem is corrected properly the first time around, you won't have added any costs due to "additional work" the technician had to do because the machine was opened. A VCR that must be returned to the service shop time and again is often the result of a communications problem between you and the repair technician.

Some repair technicians and service centers frown on consumers repairing their own VCRs. But if you've followed the directions in this book and in the manufacturer's literature and used the right tools and cleaning supplies, you've done nothing that the service shop would not have done—and charged you handsomely for in the process. When performed properly, routine maintenance and troubleshooting procedures do not harm the VCR.

Avoid a repair shop that adds an additional service fee simply because you've opened the top cover and made some preliminary tests. Unless you have broken the deck even more, the extra service fee is completely unnecessary and very unethical. After all, an automotive garage would not add $50 to their price of a tune-up because you changed the oil.

If you are sure that the fault lies in a component that cannot be repaired, and replacement does not require critical adjustment, contact the manufacturer and ask for a parts list (if you don't already have it). Identify the faulty component and order a replacement. Most manufacturers that sell replacement parts to the public ask for payment in advance, and the easiest way to do this is by credit card. Few will ship COD.

Assume the manufacturer proves to be uncooperative—what then? Try to get the part from a repair center authorized to work on your brand of player (in this case, *authorized* simply means that the repair center has an open channel for replacement parts). The shop might tack on a small service fee for obtaining the part(s) for you, which is understandable.

Receiving the replacement parts from the manufacturer or a repair center might take as little as a week or as long as several months. At worst, the broken part must be ordered from some warehouse in Japan, and it seems as if they ship it to the United States by sailboat. You've heard of the slow boat to China; in the case of replacement parts for VCRs, it's the very slow sailboat from Japan. Nevertheless, owning an American- or European-made player is no guarantee that parts will arrive any sooner.

Keep records of when you ordered the part and whether you've written any follow-up letters or phoned the manufacturer directly. Ask for the names of everyone you speak with and write them down. You never know when this additional information will come in handy.

## Finding a trustworthy repair technician

They say that once you find a good mechanic for your car, never let him or her go. The same applies to VCR repair technicians. Truly good VCR repair technicians are actually hard to find, because VCR repair shops are not all that common. Because it is often cheaper to just buy a new low-end VCR rather than have one repaired, many consumers don't bother with repairing their decks. Although the business of VCR repair is healthy, the low cost of replacement units keeps it at a moderate level.

Finding a good VCR repair technician requires some sleuthing, but it can be done. Here are some tips to get you going in the right direction:

☐ Whenever possible, ask for referrals for a good repair technician (and repair shop). Friends, business associates, video rental stores, and even home electronics stores are great places for recommendations.

- [ ] Check how long the business has been in operation. Stay away from repair shops that have been in the area for less than a couple of years.
- [ ] Always call the Better Business Bureau in your area before taking a VCR to a repair shop. They'll tell you if the business has a satisfactory or unsatisfactory rating.
- [ ] Once you locate a candidate shop, call first to make sure the business has not closed or moved.
- [ ] When you visit the shop, keep an eye out for the little things. Does the repair technician handle the front office and all the paper work? Although this might seem like the shop is a small operation, it can have its up-side too: the repair technician is more intimately involved with customers, and you might get better service.
- [ ] Always, always get estimates in writing. If possible, ask for the return of faulty parts.
- [ ] Many repair shops and technicians charge for an estimate; others do not. Don't judge a shop or repair technician because they do or don't charge for an estimate. However, stay away from a shop that charges an excessive fee for an estimate. There is no need for you to pay lots of money to possibly find out the VCR cannot be salvaged.
- [ ] After the VCR has been repaired, check its operation while in the shop. Don't sign for the VCR until you are satisfied that the problem has been corrected, and that the VCR doesn't have a new problem.

## Troubleshooting techniques

Effective troubleshooting depends mostly on common sense, but here are some tips you'll want to remember.

### Write notes

Write copious notes. Write everything down including how you removed the top cover, volt-ohmmeter readings you made, visual observations, parts you replaced—in short, anything and everything. By keeping notes, you can not only retrace your steps should you get hung up on a particular fault, but you'll be able to better deal with recurring problems. The maintenance log in chapter 5 provides blanks for notes; use additional sheets of blank paper if necessary.

## Use the proper tools

Refer to chapter 4 for more information on the proper tools to maintain and service your VCR. Don't make do with a tool that was not designed for the job. If you don't have the required tools and supplies already, spend a little extra on them. The maintenance and troubleshooting procedures outlined in this book require only the most basic hand tools and test equipment. Expensive items like oscilloscopes, function generators, and frequency counters are not absolutely required unless you opt for more detailed troubleshooting.

## Use your volt-ohmmeter correctly

A number of troubleshooting procedures require you to test the suspected component or assembly with a volt-ohmmeter. Be sure to use this piece of equipment correctly, or you could wind up missing a potential fault or replacing components that are perfectly good.

To test continuity, select the resistance function on the meter. If the meter is not autoranging, choose an initially high resistance range on the order of 10 kΩ or more. Attach the two meter leads (or probes) to either side of the switch, wire circuit, or connector you are testing, such as that shown in Fig. 8-2 below and Fig. 8-3 on the next page. Double-pole, double-throw switches, sometimes used in power switches, have extra terminals to check. Figure 8-4 on the next page shows how to use your meter to test one of these switches.

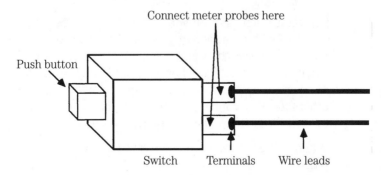

Connect meter probes here

Push button

Switch    Terminals    Wire leads

■ **8-2** *Connect the meter probes to the switch terminals to test for proper switch operation.*

Continuity is a go/no-go test. You will either get a reading of 0 Ω, which means that the circuit or connection between the two test leads is complete, or you will get a reading of infinite ohms, which means that the circuit is broken or that the connection between the two test leads is open. Sometimes, a reading of 0 Ω is exactly what

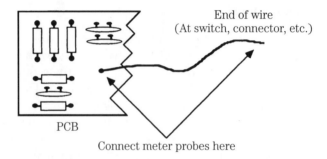

End of wire
(At switch, connector, etc.)

PCB

Connect meter probes here

■ **8-3** *Check the continuity of a wire by attaching the meter probes to both ends of the wire.*

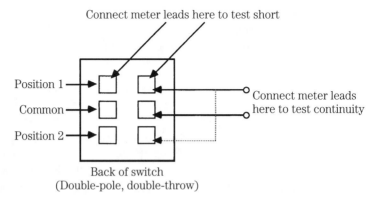

Connect meter leads here to test short

Position 1

Common

Position 2

Connect meter leads here to test continuity

Back of switch
(Double-pole, double-throw)

■ **8-4** *The correct test points when checking a double-pole, double-throw switch.*

you want, such as when you are testing the closure of a switch or a length of wire to make sure it is not broken inside. Other times, zero resistance means that something is shorted out, which is an unhappy situation. Likewise, a reading of infinite ohms (which meters display in a variety of ways), might be correct for a given test; 0 Ω for another test. The troubleshooting charts in chapter 9 provide more details on the typical readings you should get for any given instance.

When using the meter, make sure that you do not touch the metal part of the probes. Your body has a natural resistance, however high, and you will add or subtract it to the measurement you're trying to take. Always take your readings by grasping the test leads by the plastic insulator.

Volt-ohmmeters test more than continuity, of course. For example, you will want to use the ac and dc voltage functions to test for proper voltage levels going into and out of the deck power supply. Be sure to select the correct function before you connect the test leads to the live circuit. With many meters, connecting the leads to a dc voltage when the unit is dialed to ac can burn out a fuse or cause internal damage.

If the meter is not the autoranging type, always choose a range higher than the input voltage. If you are not sure of the voltage level, choose the highest one first, then work down. Almost all digital meters have built-in overload protection to prevent damage by choosing too low of a range. However, the needle in an analog meter can be permanently bent or broken if you accidentally choose a range that's too low, and the needle violently *pegs* (swings to the end of the scale).

## Going beyond the flowcharts

Few videocassette recorders are designed exactly alike. Differences between brands, even between models of the same brand, require slightly different troubleshooting procedures. Use the flowcharts in chapter 9 as a starting point only. The design and construction of your VCR might require additional troubleshooting steps. If possible, refer to the schematic or service manual of your machine, which often details testing points or procedures that are unique to the particular model.

In other cases, the design of your VCR might not require that you perform some troubleshooting procedure. For instance, if your deck is a top-loading type, it lacks a motorized tape-loading mechanism, so problems with this assembly would be of no concern to you. Likewise, camcorders lack any type of internal power transformer, so it is not necessary to test it if you are having trouble with powering your unit (except for the battery or ac adapter).

To keep the charts as simple as possible, they are limited to reflect problems inherent in front- and top-loading ac-operated home models. If you own a portable VCR or camcorder, allow for the design difference in your troubleshooting procedures. Keep in mind that portable decks and camcorders are considerably more difficult to work with because of the miniaturization of scale.

# Troubleshooting VCR malfunctions

VCRS ARE COMPLEX ELECTRONIC AND MECHANICAL DEVICES. Their reliance on special proprietary integrated circuits and zero-tolerance mechanical alignment leave the average consumer or electronics hobbyist with limited opportunity for complete home repair.

This situation does not mean, however, that doctoring the ills that beset the average VCR is entirely out of your hands. On the contrary, the new modularity of VCR design makes mechanical and electronic faults very rare—assuming that the machine was well manufactured in the first place.

In this chapter, you'll learn what you can do when your VCR goes on the fritz and how to go about fixing it. The emphasis is on troubleshooting the mechanical sections of the deck, and plenty of flowcharts are provided to show graphically the step-by-step process of the cure for a given ailment.

## Repair it yourself

By far, the most cited problems with videocassette recorders are not internal problems with a transistor or IC chip or even a broken video head. Most of the problems are relatively mundane mechanical faults caused by failure to follow the operating instructions, careless use or abuse, tape defects or damage, clogged (or dirty) video heads, and just normal wear.

More specifically, service shops for VCRs report that the following causes represent by far the greatest majority—more than 85 percent—of warranty and nonwarranty repair. The causes are listed from most common to least common.

☐ Poor system connection (cable from VCR to TV, etc.)
☐ Improperly set controls (especially control knob)
☐ Dirty heads (video, audio, and control track)
☐ Tape problems (includes bad or old tapes)
☐ Physical abuse or accidental damage (caused by neglect or improper cleaning)
☐ Dirty, glazed, or cracked reel spindle idler tire
☐ Dirty, glazed or cracked drive belts
☐ Jammed mechanical components due to oxidation, dirt, or loss of lubrication

You can save yourself the cost of a service call, not to mention headaches and frustration, by not only following manufacturers' instructions for operating your video gear but by caring for your tapes and your deck with routine maintenance, as discussed in chapter 5.

## Service politics

Inevitably, ordinary mechanical and electronic breakdown does occur. You can repair a number of these breakdowns yourself. Most dysfunctions can be corrected by cleaning dirty electronic contacts and switches, oiling or greasing the moving parts, repairing the occasional broken or frayed wire, and replacing worn parts like rollers, belts, and plastic gears.

Unless you have specific training and experience in working with both high-speed digital and analog circuits and have the proper servicing tools, problems with the video heads (other than dirt), as well as failed integrated circuits and other components on the circuit boards, should be referred to a repair technician.

More complex problems require a schematic or service manual, an oscilloscope, and a variety of specially made alignment and test jigs. They also require additional troubleshooting and repair techniques, which are beyond the scope of this book. If you are interested in learning more on this subject, see appendix B for a selected list of general and specific electronic troubleshooting and repair guides.

Many manufacturers will sell you a copy of the schematic or service manual, but be prepared to spend up to $30 for it. The wait can take up to eight weeks, so order the manual before your deck breaks down. Refer to appendix A for the names and addresses of most VCR manufacturers. If you don't see the manufacturer of your VCR listed or the manufacturer has moved, refer to the owner's manual or warranty card. It should list the main address of the manufacturer, plus local and regional repair centers.

Most VCR manufacturers will not sell the test jigs and alignment tools directly to consumers. In fact, many VCR manufacturers won't even sell the tools to independent service centers! Other than routine cleaning and checkout, along with replacement of some mechanical parts (most or all of which you can do yourself), all but "authorized" service centers are directed to send defective decks back to the manufacturer for repair.

One general exception to this rule is defects in the main printed circuit board (PCB), as well as the modular sections for RF modulator, tuner, and clock/timer. These problems can often be handled in the repair shop, but they almost always entail completely exchanging the old board for a new one. In the service trade, a technician who repairs electronic products simply by exchanging a PCB is euphemistically called a "board swapper." Technicians don't much like it (the implications in the phrase as well as the inability to get parts), but individual components and *IC*s (integrated circuits) are not always available from the manufacturer, and they must take what they can get.

If you suspect a problem with the PCB in your deck, you might be able to get a replacement directly from the manufacturer and change the board yourself. Getting a replacement yourself could save you $75 or more in labor costs. VCR makers sell the boards on an exchange basis. You send in the old, defective board—along with a check or money order to cover the service fee—and they send you a good, tested board in return. Although the replacement cost for a board varies, it is typically in the $75 to $125 range.

## Maintenance procedures and special adjustments

Most of the troubleshooting steps described in the flowcharts that follow are standard maintenance procedures covered in previous chapters. Specifically, the procedures call for routine video and audio head cleaning, replacement or rejuvenation of belts, tires, and other rubber parts, and general cleaning and lubrication. These points are more fully discussed in chapter 5, "General cleaning and preventive maintenance," and in chapter 6, "VCR first aid."

Some repair procedures and adjustments require special techniques. You might want to attempt these repairs yourself, or refer to a qualified VCR technician. Be aware that incorrect repair or adjustment can worsen the damage of your VCR, so you should only attempt service when you feel competent in the procedure and have the proper tools. In addition, do not attempt the following adjustments if there is nothing wrong with your VCR—in other words, don't fix it if it's not broken. These adjustments and repair procedures are not part of routine maintenance.

### Tape-guide spindle adjustment

In VHS decks, two tape-guide spindles (or posts) draw the tape tight against the spinning video heads. These guides are mounted in a slotted track. The guides, shown in Fig. 9-1, serve two purposes: To route the tape around the head drum at the proper angle, and to provide the correct tape height for the video-head drum for proper recording and playback.

■ **9-1** *One of two tape-guide spindles in a VHS deck.*

Although the track usually requires cleaning and relubrication every 12 to 18 months, the guides seldom need adjustment. However, thermal expansion and contraction (caused by normal operation of the VCR) can make the guides loose and out of adjustment.

When the guides are out of adjustment, the deck loses partial or complete tracking ability, and using the tracking control has no effect or limited effect. Newly recorded tapes made with the guides out of whack might look fine, but they might not be playable on a properly adjusted deck. What's more, subsequent adjustment of the guides in your VCR will render the previously made tapes unwatchable.

The height of the guides can be adjusted by loosening a set screw (usually at the bottom of the guide) and then turning the adjustment screw on top. The head of the adjustment screw is either a special slotted head or a hex (typically 1.5 mm). If your deck has a slotted-head adjustment screw, you need to make a special tool by filing down the center of a flat-blade screwdriver, as shown in Fig. 9-2.

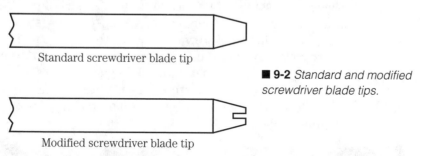

Standard screwdriver blade tip

■ **9-2** *Standard and modified screwdriver blade tips.*

Modified screwdriver blade tip

In a service center, the adjustment of the tape-guide spindles is made by connecting an oscilloscope to a test point within the VCR and then adjusting the height of the guides until the waveform on the scope is at its maximum peak. You can duplicate this method if you have an oscilloscope and the proper service manual for your deck (the manual should list the test point and what the waveform should look like).

Lacking these items, you can watch the picture on a good TV while playing back a commercially prerecorded movie (or use a test tape). Center the tracking control knob on the front of the deck; then adjust the guides until all of the lines on the top and bottom of the TV screen are gone. Go slowly—a small turn makes a big difference.

Because there are two tape guides, you must adjust each one for the best picture. Write notes as you go, and if you find the picture is getting worse, turn the guide back to where it was and try the other one. Count the number of revolutions and the direction you turn the guides so you can return to where they were if necessary and start all over again.

Be extra careful when fine tuning the tapeguide spindles, and then only perform the procedure when you are certain that the tape guides are out of adjustment. Don't fiddle with the guides just because you think the picture ought to look better. If you are unsure of your abilities, take your deck to a service center and have them make the adjustment.

## Control/audio head adjustment

On all but the very old VCRs, the control-track and audio heads are combined in one unit (VHS stereo models have both right and left channels in the head unit and might have an audio-erase element as well). In service literature, this head is often referred to as the *ACE* head, for audio, control, and (audio) erase.

The linear distance of the head unit to the video head drum can be adjusted by turning one cone or V-shaped screw (shown in Fig. 9-3). By turning the screw one way or the other, the head unit moves closer to or farther away from the video-head drum, thus retarding the control-track and audio signals relative to the video information. Adjusting this screw is for tracking noise in the bottom or top of the picture (see Fig. 9-4) or to rectify loss of sound-picture synchronization.

■ **9-3** *The control/audio head adjustment screw (or cone).*

The control/audio head does not need to be adjusted unless the heads are replaced or somehow become out of whack. The adjustment should be made using an alignment tape and by following the instructions in the service manual. But if you desire, you can adjust the head distance yourself by using a commercial prerecorded movie as a test tape and adjusting the screw until the audio signal sounds clear and the picture is sharp.

■ **9-4** *The effect of a maladjusted control/audio head screw is similar to mistracking, and it breaks up the picture into lines.*

## Erase-head tape post adjustment

On most VCRs, either the full-erase head or the tape guide post immediately before the full-erase head is adjustable. The height of the head or guide should be checked if the full-erase head is replaced. You can always tell that the full-erase head is not completely erasing the tape if you record a new program over an old one. The video tracks will almost always be completely erased, but remnants of the control and audio tracks might remain. A breakup of the picture can mean that the old control track is not being completely erased. In addition, background sound or a loss of overall sound fidelity can mean that the audio tracks are not being erased.

The erase head or post is adjusted in a similar manner to the tape-guide spindles and control/audio head, as explained above. As before, go slowly and note the results after twisting the set screw or post ¼ or ⅛ turn.

## Other adjustments

Depending on the age and make of your VCR, there are other adjustments that you can make to repair your deck or make it run more smoothly. Most of the following adjustments are not found on VCRs made after 1982 or 1983. Check the service manual for your particular machine for the adjustment procedure:

☐ Cassette-in leaf-switch adjustment. Adjusts the switch for active contact, as shown in Fig. 9-5, when a cassette is inserted in the machine.

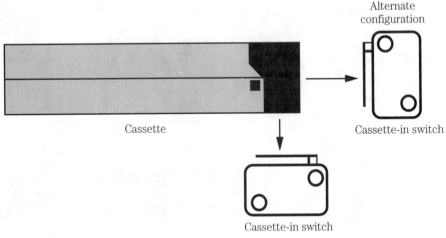

Cassette

Alternate configuration

Cassette-in switch

Cassette-in switch

■ **9-5** *Two possible configurations for the cassette-in switch.*

☐ Loading-end leaf-switch adjustment. Adjusts the switch for active contact when the loading cycle is complete and the tape-guide spindles are seated against the stops.

☐ Back-tension tape adjustment. Alters the degree of back tension on the supply-reel brake, as determined by the back-tension guide or lever.

☐ Supply and take-up reel torque adjustment. Adjusts the amount of torque available for REWIND, FAST FORWARD, and PLAY operations at the supply and take-up reel or spindles.

☐ Capstan pinchroller adjustment. Adjusts the amount of pressure exerted by the capstan pinchroller.

☐ Capstan free-running speed. Adjusts the free-running speed of the capstan. Use a frequency meter or oscilloscope for precise adjustment. Some VCRs have timing marks on the belts to determine correct capstan speed. You need a neon lamp or strobe jig to make the adjustment.

☐ Video-head switching-point adjustment. Adjusts the switching point between the two video heads during playback (affects SP and EP heads in a four-head deck, or there might be two controls for the two sets of heads in a four-head deck).

☐ Reel spindle-height adjustments. Alters the height of the supply and take-up reel spindles so they are properly aligned.

Improperly aligned reel spindles cause scalloping or mistracking. This adjustment is very crucial and should not be made without the use of a test jig and alignment tape (most reel spindles use small spacers for proper adjustment; exchange worn-out spacers with exact replacements).

# Using the troubleshooting flowcharts

The remainder of this chapter is devoted to a series of trouble-shooting flowcharts that detail the possible causes and suggested solutions for a series of common and not-so-common VCR ailments. The flowcharts are not meant to be definitive, but they should go a long way in helping you pinpoint a problem in your deck. See chapter 8 for an explanation on how to use and interpret the flowcharts. Refer to chapters 4 and 5 on the tools and supplies for VCR maintenance and repair and how to use them.

To prevent overcomplicating the flowcharts, they have been de-signed to apply specifically to ac-operated home decks. Many of the same problems, causes, and solutions apply to portable VCRs and camcorders, but troubleshooting techniques can differ in certain situations.

You can still use the troubleshooting flowcharts to diagnose most problems in portable or camcorder units, however, because the basic reasons behind the fault will be similar. For example, a camcorder does not have an ac cord, so you would not test its cord if you are having trouble with it. But you would test the battery or battery charger in a similar manner as you would an ac cord.

Another reason for leaning toward the home ac-operated units is that they are considerably easier to maintain and repair than the portable and camcorder versions. The cramped real estate in a portable VCR or camcorder means that the parts are smaller, more fragile (generally), and harder to reach. As a matter of course, it is suggested that you not service a portable VCR or camcorder unit unless you have sufficient mechanical skills to do so.

## Look to the simple things first

In using or looking over the flowcharts, some possible causes might seem overly simplistic. The biggest mistake you can make in servicing your VCR (or anything else for that matter) is overlooking the obvious. I am reminded of the time, in my younger and more foolish days, when a color television set came in for repair. The set would turn on and you could hear sound, but there was no picture. It was an old tube set and we tore into it with a vengeance, replacing every tube that could possibly cause the problem. Alas, we replaced one tube after the other, with no results.

Then, innocently enough, we began to fiddle with the control knobs. We hit upon the brightness control, and upon turning it up, the picture magically came on! After spending more than $50 on tubes (in the 1960s, 50 bucks was a lot more money than it is now) and wasting a whole day trying to fix the set, the problem turned out to be a brightness control that was inadvertently turned down.

This lesson was an expensive one, but it was a good one. It proved that if something doesn't work, odds are it's a simple cause with an equally simple solution.

## List of flowcharts

Here are the troubleshooting flowcharts in their order of presentation. The flowcharts are numbered for cross-reference. Within the flowcharts, the steps (or levels) are numbered and correspond to numbers in the text. Unless otherwise specified, all troubleshooting procedures and tests should be done with the deck turned off and unplugged (or removed from the power source).

1. VCR does not turn on
2. VCR turns on but nothing else
3. Cassette will not load
4. Cassette will not eject
5. VCR will not thread tape
6. VCR will not play tape
7. VCR eats tape
8. Fast forward or rewind won't operate
9. Search does not work correctly
10. Tracking control has no effect
11. VCR does not respond to some or all front-panel controls
12. Front-panel indicators not functioning
13. Timer does not operate properly
14. Tuner channels don't change on TV
15. Sound okay; video not okay
16. Video okay; sound not okay
17. Snowy video; poor audio
18. Remote control does not operate or function properly

19. VCR makes unusual mechanical noises during loading and playback

20. You receive an electrical shock when you touch the VCR

21. VCR overheats

At the end of this chapter is a set of troubleshooting guidelines for repairing miscellaneous VCR malfunctions.

## Flowchart 1. VCR does not turn on

The VCR is dead—nothing, even kind words, seems to revive it. Pushing the power switch has no effect, and the front-panel controls, even the clock and timer, are inoperative. Here is what to look for.

### Step 1-1

If the deck runs on ac power, the first logical step is to ensure that it is getting power from the ac wall source. Make sure that the cord is plugged in and that the polarized plug is inserted properly. Do not defeat the purpose of the polarized plug by filing or cutting the wide prong.

Most homes are equipped with switched outlets—you flick a wall switch to turn the outlet on and off. Sometimes the switch controls both outlet receptacles, and other times it only controls one. If the VCR is plugged into a switched outlet, make sure that the outlet is turned on.

To rule out the possibility that the problem originates in the wall outlet and not in the deck, plug a lamp into the socket and see if the lamp works. If the light glows, obviously the outlet is good. If the lamp seems dim, carefully check the voltage at the outlet with a volt-ohmmeter. It should read between 108 and 125 Vac (117 is average). Be absolutely sure that you follow all safety precautions when testing the ac voltage or you can receive a serious shock.

### Step 1-2

VCRs with timers usually shut down when in the TIMER mode (the timer starts recording at a specific time of the day). Pressing the POWER switch might do nothing, and the deck could act as if it is inoperative. Follow the manufacturers instructions on deactivating or resetting the timer.

1. VCR does not turn on

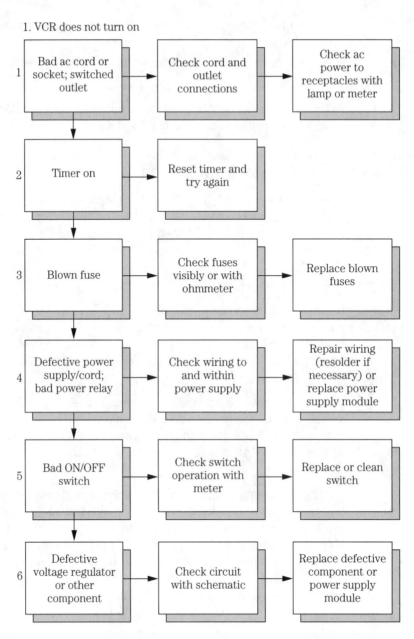

| | | |
|---|---|---|
| 1. Bad ac cord or socket; switched outlet | Check cord and outlet connections | Check ac power to receptacles with lamp or meter |
| 2. Timer on | Reset timer and try again | |
| 3. Blown fuse | Check fuses visibly or with ohmmeter | Replace blown fuses |
| 4. Defective power supply/cord; bad power relay | Check wiring to and within power supply | Repair wiring (resolder if necessary) or replace power supply module |
| 5. Bad ON/OFF switch | Check switch operation with meter | Replace or clean switch |
| 6. Defective voltage regulator or other component | Check circuit with schematic | Replace defective component or power supply module |

## Step 1-3

A blown fuse will prevent your VCR from operating. If your deck is equipped with an external fuse, remove it and visually inspect it. If you can't tell that the fuse has blown, use a meter to check continuity. Attach the leads to either side of the fuse. If the fuse is good, the meter will read 0 Ω. Anything else indicates a bad fuse.

Many fuses are located inside the deck, on or near the power supply section, as shown in Fig. 9-6. You must remove the top cover of the VCR to inspect the internal fuses. Some fuses are soldered in place but can be tested using a volt-ohmmeter. You should get a reading of 0 Ω when the test leads contact either side of the fuse. Be absolutely sure that the VCR is off and unplugged when testing fuses.

If the fuse is bad, use an exact replacement only. If the value of the fuse is 0.375 A (amperes), do not use a 1 A fuse.

■ **9-6** *An internal fuse, located near the transformer in the power supply section.*

### Step 1-4

After checking the ac sources, timer, and fuse(s), inspect all power cords for cracks and other signs of wear. A badly worn or damaged cord can cause a number of problems. A short will blow fuses (either in the house or in the deck); an open line will prevent the ac from reaching the deck.

Check the power cord using a meter. An open will register infinite ohms when the test leads are connected across the two prongs of the ac cord. A short will register 0 Ω when the test leads are connected to the prongs and corresponding internal wiring of the deck, as shown in Fig. 9-7. Test both prongs.

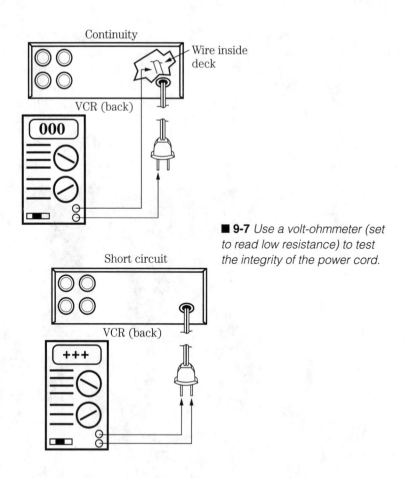

Continuity

Wire inside deck

VCR (back)

000

**■ 9-7** *Use a volt-ohmmeter (set to read low resistance) to test the integrity of the power cord.*

Short circuit

VCR (back)

+++

If readings are erroneous, test the continuity of both the coil and the relay terminals. The coil should remain continuous, but the contact terminals should be shorted when on and open when off (see Fig. 9-8).

If the wiring to the switch (or relay) tests okay, try the wiring leading that goes from the switch to the inputs of the power transformer (called the *primary*). Also test the wiring from the output of the transformer (called the *secondary*) to the power supply section or main printed circuit board. A typical schematic diagram of the power supply circuits for a typical VCR is shown in Fig. 9-9. Note there are many different voltages generated by the power supply.

To test whether the transformer is delivering power, plug in the machine (if it is safe) and turn it on. With the meter set to read ac volts, in a range of no less than 25 or 50 V, connect the test leads to the outputs (secondary) of the transformer. Without a schematic, it

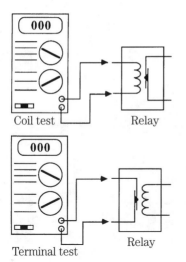

■ **9-8** *A meter can be used to test the coil and terminals of the relay. The coil test can be performed with the power on, but be sure to use the proper meter settings and range.*

might be difficult to make sense of the reading, but generally you should get a reading of 6 to 24 V when connecting the leads to any of the wires.

The power transformers used in most VCRs have more than one tap-off point, so applying the leads at various tap-off terminals yields a variety of voltage levels. You can be fairly sure the transformer is working properly if you get some readings when the meter is connected to most of the secondary wires. Be sure to avoid the primary wires, as they carry the full 117 Vac from the wall outlet.

## Step 1-5

The power wiring inside your deck could be faulty. Open the VCR and check the connections leading to the power switch. Check for shorts and open circuits using the meter. Normally, only one side of the incoming ac is connected to the switch. The other side should connect directly to the power transformer.

If the deck has a relay, check it to make sure that it is getting power. The relay is usually located near the power transformer; check the heavy wires (the two smaller wires lead to the power switch and carry low voltage only). Plug the VCR in and set the meter to ac volts. Connect one test lead of the voltmeter to ground and the other to the incoming wire going to the relay. There should be power. Now, attach the lead to the other power terminal on the relay. With the relay off, there should be no power; turn the VCR power switch on and the meter should register approximately 117 Vac.

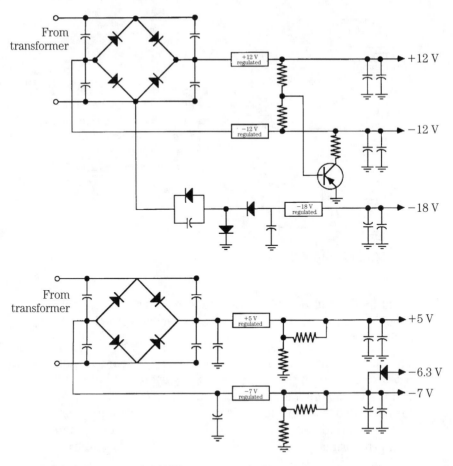

■ **9-9** *A representative VCR power supply. Note the supply provides several voltages.*

### Step 1-6

If you get a reading in the transformer secondaries but the VCR still appears inactive, there is probably a problem with the power supply board. A burned-out voltage regulator will cause partial or complete loss of function; similar results will happen if a capacitor or other power supply component is shorted. Visually inspect the power supply section and look for charred components. Look especially at the voltage regulators (they look like power transistors) and the large filtering capacitors.

A defective voltage regulator might be hard to spot visually but can be checked using a voltmeter, as shown in Fig. 9-10. For best results, refer to a schematic to determine the proper test points. The regulator could have several pins, and the pins might not be

■ **9-10** *Use the volt-ohmmeter carefully to test the operation of the voltage regulator (attached to the opposite side of the board).*

identified on the PCB. Black charring or an oozing, dark substance around a large electrolytic capacitor indicates trouble. Replace the capacitor with the same value (both in microfarads and volts) or replace the power supply module as a whole.

## Flowchart 2. VCR turns on but nothing else

It is often harder to troubleshoot a VCR when the machine turns on but doesn't play; however, at least you can rule out problems in the power supply, fuses, on/off switch, and the relay.

### Step 2-1

Taking a VCR from the cold, damp outside air to the warm, inside air can cause condensation to form on the internal components. Many VCRs (including portables and camcorders) have a dew sensor that prevents the machine from operating when there is condensation present. If you suspect condensation, see if the dew indicator is on (some decks don't have a dew indicator, but the STOP button indicator might flash instead). Also, visually inspect for moisture and wait 30 to 60 minutes for it to evaporate. You can speed up the drying process by using a hair dryer. Keep the dryer on low heat or no heat.

2. VCR turns on but nothing else

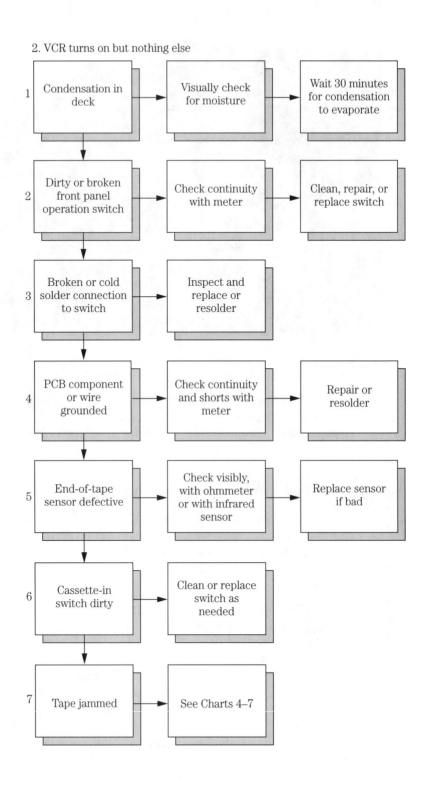

| | | | |
|---|---|---|---|
| 1 | Condensation in deck | Visually check for moisture | Wait 30 minutes for condensation to evaporate |
| 2 | Dirty or broken front panel operation switch | Check continuity with meter | Clean, repair, or replace switch |
| 3 | Broken or cold solder connection to switch | Inspect and replace or resolder | |
| 4 | PCB component or wire grounded | Check continuity and shorts with meter | Repair or resolder |
| 5 | End-of-tape sensor defective | Check visibly, with ohmmeter or with infrared sensor | Replace sensor if bad |
| 6 | Cassette-in switch dirty | Clean or replace switch as needed | |
| 7 | Tape jammed | See Charts 4–7 | |

If the moisture does not evaporate, inspect the dew sensor. One is shown in Fig. 9-11. Its location in the deck varies, but it is always in the tape-transport area. Wipe off the sensor with a clean cloth or clean it with pure alcohol (the alcohol will displace the water, speeding up the drying process). Check the leads connecting the sensor to the deck. If they are shorted, the VCR will assume moisture is present and shut the machine off.

■ **9-11** *The dew sensor.*

## Step 2-2

You might not be able to control the operation of the deck if one or more of the front-panel switches are dirty or broken. If only one switch is affected, you might be able to cycle the deck through its other operations. You can isolate that switch by finding its solder contacts or connecting wires on the switch panel PCB (the switch panel might be a part of the main PCB in some decks). Use the volt-ohmmeter to test the continuity of the switch. Pushing the switch should change the reading from zero to infinite ohms or vice versa.

## Step 2-3

If all the front-panel switches are inoperative, the problem could be in the common connecting wire that goes to the switches (if applicable) or to the front-panel PCB. Test all the wires leading to the switch panel with the meter to make sure none are broken or loose. With the test leads connected to either side of the wire, a reading other than 0 Ω is an indication of an internal break in the wire. Inspect the solder points for signs of a cold or incomplete solder joint. Resolder the joint if necessary.

With many decks, the front-panel PCB is attached to the main PCB by connectors, as shown in Fig. 9-12. Use your meter to test the continuity between the connectors. If you find a reading other than 0 Ω, carefully remove the connector and inspect it for loose wires, broken wires, and broken or dirty contacts. Resolder or replace the connector if it is damaged. Use a recommended cleaner to clean contact points.

■ **9-12** *Connector on the main printed circuit board.*

## Step 2-4

If the problem still persists, the next logical step is to examine the wires and components on the main PCB. Are there any obviously broken wires? Test with a meter to be sure. Look closely at the capacitors and resistors mounted on the board. Occasionally, you can spot a burned-out component by looking for black charring on and around it.

Generally, you cannot spot a blown transistor or integrated circuit in this manner because the damage does not usually extend to the exterior of the device because of the low operating voltages of these devices. Refer to the service manual for the proper voltage levels at the pins of the transistors and ICs. Use your meter to check the voltages.

## Step 2-5

In VHS decks, the end-of-tape lamp is used to indicate when the tape has reached the beginning or end. The tape normally blocks the light from the lamp to the two sensors located on either side of the transport area. But when the tape reaches the beginning or end, the light shines through the clear leader portion and strikes against a light-sensitive transistor—the sensor. When the sensor receives light, playback stops and the tape unthreads.

If the end-of-tape lamp is out (it is usually infrared so you can't see it), the deck will still operate but the safety feature of the automatic shutoff is lost. That should be avoided to preserve the tape. Test the lamp with the infrared sensor described in chapter 4. If either of the sensors is shorted or defective or its connecting wires are shorted, the deck might assume that the end of the tape has been reached and will place itself in the auto shutoff mode.

259

Inspect the sensors to be sure that they are not shorted. Use your meter to measure the output voltage of each sensor (set the meter to dc volts). Block the sensor from light and the reading should be fairly low. Shine a light at the sensor and it should go up to about 5 to 12 V. If the voltage of the sensor stays low or high no matter how much light strikes it, inspect the connecting wires for shorts. The sensor itself might be defective; replace it if so.

Beta decks do not use light to detect the end of the tape. Rather, they use electromagnetic coils that detect the absence or presence of ferrous material. Two coils are used, as explained in chapter 2, that constantly detect the ferrous surface of the tape. At each end of the tape is a small sliver of aluminum foil. When the coil touches the foil, the VCR stops playback.

You can test the coils in the same manner as the VHS lamp sensors. Simulate a tape by placing a washer or other ferrous object on or near the coil. Watch the voltage on the meter. Now remove the object and watch for a voltage change. If the voltage doesn't change, look for shorts and inspect and test the coil. Trace the leads from the coil back to the PCB to make sure there are no breaks.

*Flowchart 2. VCR turns on but nothing else*

### Step 2-6

All VCRs have a cassette-in switch, located on or near the cassette lift mechanism. If this switch is not activated, the VCR assumes that no tape is inserted, and none of the operator controls will function. Inspect your deck and find the switch. If the switch can't be found, refer to the service manual. When the cassette is inserted and loaded (either manually with a top-loading VCR or automatically with a front-loading unit), is the switch activated? Use your meter to check for continuity of the switch.

### Step 2-7

If there is a tape in the deck, the tape itself could be jammed and preventing playback. See Flowcharts 5, 6, and 7 for more information on locating and servicing tape jams.

## Flowchart 3. Cassette will not load

If you find that the cassette will not load in the machine, under no circumstances should you try to force the tape. Doing so can seriously damage the VCR and will surely wreak havoc on the tape. Most of the steps in this flowchart assume a front-loading VCR.

### Step 3-1

Is the power on? This question might seem insulting, but remember it's the obvious things that are often overlooked. Also check to see if the deck is in the TIMER mode (waiting to record a show). Most decks "play dead" when in the TIMER mode; switch it to the normal mode to insert the tape.

If the power is on and the timer is off, try pushing the ON/OFF button a few times to force the deck to go into EJECT mode (note: some VHS machines don't do this). The loading and threading mechanism might not have been initialized at their starting positions, and cycling the deck could put everything back to where it should be.

### Step 3-2

If the VCR already has a tape inside it, the machine certainly won't take another. Many front-loading VCRs use a smoked-plastic front panel over the loading slot, making it impossible to tell if a tape is already inserted. However, most decks have a TAPE indicator that glows when a tape is inside. Don't blindly trust the indicator; it could be burned out or the sensor switch inside the deck that detects the tape could be dirty or broken.

3. Cassette will not load

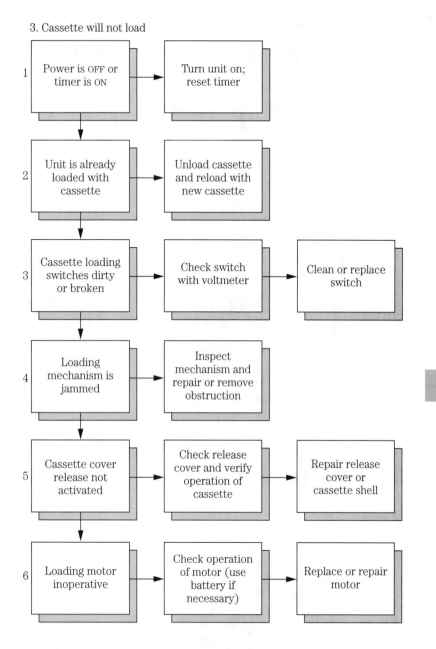

| 1 | Power is OFF or timer is ON | → | Turn unit on; reset timer | | | |
| 2 | Unit is already loaded with cassette | → | Unload cassette and reload with new cassette | | | |
| 3 | Cassette loading switches dirty or broken | → | Check switch with voltmeter | → | Clean or replace switch |
| 4 | Loading mechanism is jammed | → | Inspect mechanism and repair or remove obstruction | | | |
| 5 | Cassette cover release not activated | → | Check release cover and verify operation of cassette | → | Repair release cover or cassette shell |
| 6 | Loading motor inoperative | → | Check operation of motor (use battery if necessary) | → | Replace or repair motor |

261

## Step 3-3

If the tape does not load and thread when you insert the cassette through the loading slot, it might not be receiving the command to do so. Figure 9-13 shows a schematic diagram of the cassette-in switch that detects the presence of the tape. Note that the loading motor will not operate unless the cassette triggers the internal leaf

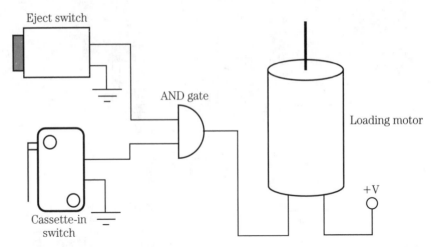

Eject switch

AND gate

Loading motor

Cassette-in
switch

+V

■ **9-13** *The loading motor and loading logic. The AND gate delivers power to the motor only if both the cassette-in switch and eject switches are on.*

switch. This cassette-in switch also is used to initiate threading— no switch contact, no threading.

Make sure that the leaf switch hasn't gotten dirty, broken, or corroded. To test the switch, connect the test leads of the meter to either side of the switch terminals. The meter should alternate between zero and infinite ohms when the switch is pushed.

Also keep an eye out for cracked or pinched wires leading to the switch, as well as poor solder joints. Test the joints and wires using the meter, as usual. Opens and shorts are indications that the wiring is bad and must be repaired.

## Step 3-4

When you insert the tape, does the loading motor operate? If it does but nothing else happens, suspect the belt (if any) that connects the motor to the loading mechanism (in some machines, the capstan operates loading; a belt connects the capstan to the loading mechanism).

Belts can break or slip off the metal or plastic pulleys. Look at the belt as you insert the tape. If the motor shaft turns, but the belt doesn't budge, the belt might be too loose or slippery from oil, grease, or wear. Inspect the belt and clean it if necessary.

A number of machines use plastic or metal gears to drive the loading (or cassette-lift) mechanism from the motor. A gear is mounted on the motor shaft and meshes with another gear or a straight-tooth rail. Inspect the gears to be sure that none of the teeth are broken and that the gear teeth are meshing properly. Even if the loading mechanism looks fine, the gears can still bind against one another. Would oiling or greasing help? Apply a small dab of grease on the gear surfaces or oil to shafts and levers.

If the tape still won't load, inspect the path of the mechanism including all gears, cams, tracks, and other parts in and around the loader. Carefully remove any objects that might have lodged in place.

## Step 3-5

Cassettes have a release button that opens the tape cover. If this cover doesn't open, the deck can't grab the tape to thread it. The cover might not open for two reasons: the cassette itself is damaged, so pushing the release button has no effect, or the release catch in the VCR is broken or bent. Check the release catch (it's on the left of the loading mechanism for Beta and on the right for VHS) and visually see if the pin is engaging with the catch on the cassette, as shown in Fig. 9-14. Note that there is no release catch for VHS-C camcorders and that the catch is on the left for 8mm cassettes.

If the cassette is bad, you might be able to repair it using an empty shell of another discarded cassette. Transfer the tape from the bad shell to the good shell, as discussed in chapter 5. If the release catch is bent, use a pair of pliers to bend it back into shape. The catch might need to be replaced if it is seriously bent or broken.

■ **9-14** *The cover release pin, located on the side of the cassette-loading mechanism.*

## Step 3-6

To ensure that the proper drive signals are being applied to the loading motor when the cassette-in switch is depressed, connect the ohmmeter to both motor terminals (see Fig. 9-15). Select the dc volts function on the meter with a range of no less than 12 V. With the deck plugged in and turned on, depress the EJECT button once (or insert a tape). With most machines, your meter should read a voltage (it can be as small as 3 V or as high as 12 V).

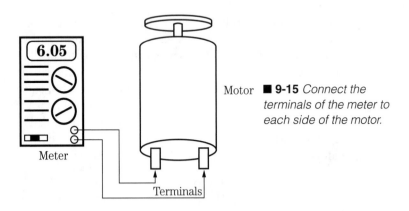

■ **9-15** *Connect the terminals of the meter to each side of the motor.*

If no other defects can be found up to this point, the problem could lie in the loading motor itself. You can test the motor by connecting a C- or D-size battery to the motor terminals, as shown in Fig. 9-16. The polarity of the battery determines which direction the motor turns, so if the motor seems to labor when power is applied in one polarity, stop and try the other direction.

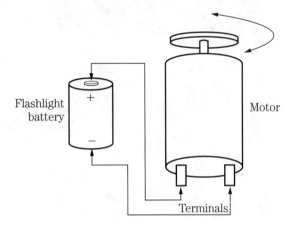

■ **9-16** *Test the operation of small dc motors with a 1.5 V flashlight battery.*

# Flowchart 4. Cassette will not eject

Many of the culprits responsible for the cassette not loading in a front-loading VCR also are responsible for the cassette not ejecting. Again, remember that under no circumstances should you try to force the tape. Doing so can seriously damage the VCR and will surely wreak havoc on the tape.

4. Cassette will not eject

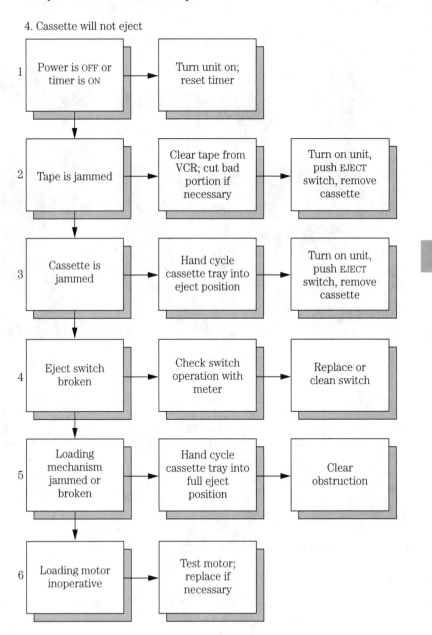

265

## Step 4-1

Refer to the previous flowchart on power and timer operations. Force the eject cycle by turning the deck off and on (once should do it). Manually inspect the loading and threading mechanism to see if it is caught in midcycle. If it is and switching the power on and off doesn't do it, manually rotate the loading or threading mechanism to bring it back to the starting point. Figure 9-17 shows the underside of a VHS deck and how to roll the threading mechanism manually back to the unthreaded position.

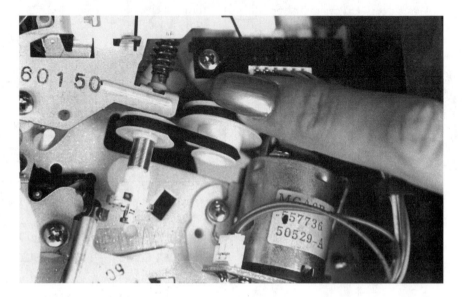

■ **9-17** *Rotate the loading motor shaft or pulley manually to engage or disengage the tape-loading mechanism.*

## Step 4-2

Inspect the inside of the VCR for spilled tape. If it has spilled into the innards of the deck, the VCR won't be able to eject the tape fully. Manually extract the tape to eject the cassette. By hand, cycle the threading and cassette-lift mechanisms to their starting positions (see previous section), or turn the deck off and on. Never force the threading mechanism by hand by pushing on the tape-guide spindles. The same applies to the loading mechanism: never pull up on the cassette carrier.

## Step 4-3

Even if no spilled tape is evident, the cassette might be stuck in the cassette-lift mechanism. Look for obstructions or blockage. Attempt to manually cycle the mechanism but do not force it. If the cassette-lift mechanism is in the up position (the cassette is not ready for threading), you might be able to simply push the cassette out with your fingers. Again, do not force it or you can cause considerable damage.

## Step 4-4

The EJECT switch could be inoperable. Make sure that the switch hasn't gotten dirty, broken, or corroded. To test the switch, connect the test leads of the meter to either side of the switch terminals. The meter should alternate between zero and infinite ohms when the switch is pushed. Keep an eye out for cracked or pinched wires leading to the switch, as well as poor solder joints. Test the joints and wires using the meter, as usual. Test the wires that lead from the external EJECT switch to the PCB or loading motor. Inspect the wires carefully and look for obvious damage.

## Step 4-5

Obstructions in the cassette-lift mechanism can prevent cassette ejection. When you press the EJECT switch, does the loading motor operate? If it does but nothing happens, suspect the belt (if any) that connects the motor to the loading mechanism. Belts can break or slip off the metal or plastic pulleys. Watch the belt as you push the EJECT button. If the motor shaft turns but not the belt, the belt could be too loose or slippery from oil or grease. Inspect the belt and clean or replace it.

Many VCRs use the same motor to operate the cassette-lift mechanism (front-loaders only) and the threading mechanism. A cam is used to cycle between the two. When a cassette is inserted, the motor first operates the cassette-lift mechanism and drops the tape downward. The cam then operates the threading mechanism, threading the tape around the heads. Inspect this cam to be sure it isn't broken. If the cam is loose, it might spin on its shaft and either not work at all or cause the deck to go into the wrong cycle at the wrong time.

Cassette-lift mechanisms that use plastic or metal gears also can be jammed. Inspect the gears to be sure that none are broken and that the gear teeth are meshing properly. Look for binding caused by poor lubrication or incorrect alignment. Apply a small dab of grease on the gear surfaces or oil to shafts and levers, and try realigning any obviously bent parts.

267

If the tape still won't eject, inspect the path of the mechanism, including all gears, cams, tracks, and other parts in and around the loader. Carefully remove any objects that might have lodged in place.

### Step 4-6

It is unlikely, but the loading motor could be inoperative due to a burnout or broken wires. Check Flowchart 3 for details on how to test the loading motor for proper operation.

## Flowchart 5. VCR will not thread tape

The tape loads, but when you press PLAY, nothing happens. The VCR won't engage the tape to thread it around the heads (in Beta decks, threading is done after loading; you don't need to press the PLAY button). Check these faults when the deck will not thread the tape.

### Step 5-1

Some VCRs have a separate motor for the threading mechanism (all top-loading decks have a threading motor only; there is no motor to operate the loading mechanism). When you press the PLAY button (or insert a tape in a Beta deck), does the motor turn and nothing happen? If this is the case, check for a broken or slipping belt. The belt connects between the threading motor and the threading mechanics. Some VCRs use rollers to convey motion between motor and threader, so look for these, too. Inspect the belt or roller and clean or replace, as necessary.

On many of the newer VCRs, the threading mechanism is operated by a set of gears. These gears can bind if they run dry. Lightly lubricate the gears as discussed in chapters 4 and 5.

Many VCRs use the same motor to operate the cassette-lift mechanism and the threading mechanism. A cam is used to cycle between the two. When a cassette is inserted, the motor first operates the cassette-lift mechanism and drops the tape downward. The cam then operates the threading mechanism, threading the tape around the heads. Inspect this cam to be sure it hasn't become broken. If the cam is loose, it might spin on its shaft and either not work at all or cause the deck to go into the wrong cycle at the wrong time.

### Step 5-2

The threading mechanism, particularly on older decks, can be complicated. It could be jammed from a foreign object or be binding due to lack of lubrication. To gain access to the complete mecha-

5. VCR will not thread tape

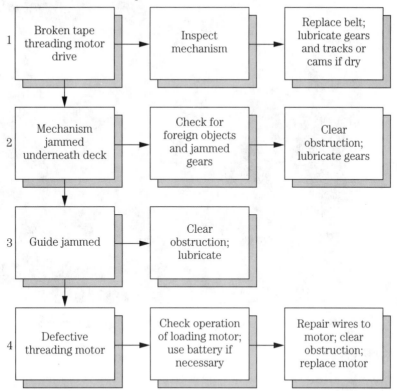

| 1 | Broken tape threading motor drive | → | Inspect mechanism | → | Replace belt; lubricate gears and tracks or cams if dry |

| 2 | Mechanism jammed underneath deck | → | Check for foreign objects and jammed gears | → | Clear obstruction; lubricate gears |

| 3 | Guide jammed | → | Clear obstruction; lubricate |

| 4 | Defective threading motor | → | Check operation of loading motor; use battery if necessary | → | Repair wires to motor; clear obstruction; replace motor |

nism, you'll probably need to remove the bottom panel. With older decks, you must pull out the entire tape transport assembly from the chassis; do this with extreme care.

Inspect the entire mechanism and cycle it by hand through complete threading (do not force). Remove the obstructions, if any, or apply lubricant to areas that seem to be binding. Remember to go lightly on the grease; too much is almost as bad as none at all.

## Step 5-3

The tape-guide spindles that grab the tape and wrap it around the head might be jammed. In VHS machines, there are two tracks on either side of the video-head drum where the guides slide. Check these tracks for obstructions. Clean with a cotton swab as shown in Fig. 9-18, and spread the lubricant to distribute it evenly.

Beta decks use a threading ring to wrap the tape around the heads. Lubricate the track where the ring slides.

**■ 9-18** *Clean the tape guide spindle track with a cotton swab.*

### Step 5-4

On rare occasions, the threading motor could be defective. You can test the motor by connecting a C- or D-size battery to the motor terminals, as shown in Fig. 9-16. The polarity of the battery determines which direction the motor turns, so if the motor seems to labor when power is applied in one polarity, stop and try the other direction.

## Flowchart 6. VCR will not play tape

If the tape loads and threads but does not play, suspect the following problems. Be especially careful if the tape threads around the video heads but doesn't move—considerable tape damage could result and the heads can become especially dirty.

### Step 6-1

Inspect the cassette to make sure it is not defective. The tape might be caught on the cover or some other component, or the internal reel locks in the shell could be preventing the tape from moving.

To test the cassette, open the cover (by pressing the release button on the side). With Beta cassettes, opening the cover automat-

6. VCR will not play tape

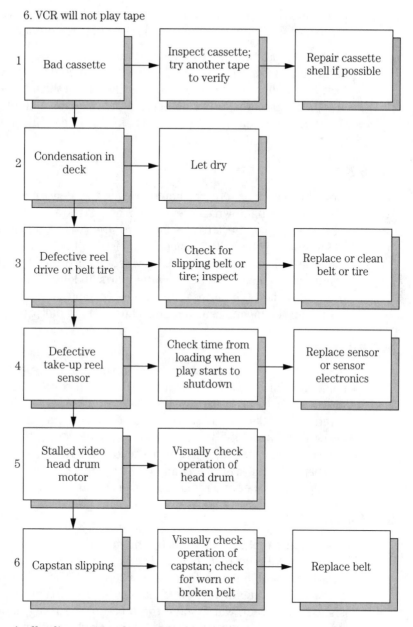

| | | |
|---|---|---|
| 1. Bad cassette | Inspect cassette; try another tape to verify | Repair cassette shell if possible |
| 2. Condensation in deck | Let dry | |
| 3. Defective reel drive or belt tire | Check for slipping belt or tire; inspect | Replace or clean belt or tire |
| 4. Defective take-up reel sensor | Check time from loading when play starts to shutdown | Replace sensor or sensor electronics |
| 5. Stalled video head drum motor | Visually check operation of head drum | |
| 6. Capstan slipping | Visually check operation of capstan; check for worn or broken belt | Replace belt |

*271*

ically disengages the reel locks, so you can now turn the reels to test the free movement of the tape. With VHS cassettes, you must depress the reel lock button, located in the center on the underside of the shell. Turn the reels toward the center of the cassette. Try both reels to make sure that the tape moves freely back and forth within the cassette. Cassettes for 8mm VCRs use a similar "release" as VHS; VHS-C cassettes have no reel locks.

If you find the tape won't move, disassemble the shell, as described in chapter 5, "General cleaning and preventive maintenance."

In some cases, the tape threads and the deck attempts to play it, but it stops after a few seconds. This problem could be caused by heavy scratches on the tape that allow light from the infrared end-of-tape lamp to reach the detector. Visually inspect the tape for scratches. Try another tape or fast forward the tape until the scratch is cleared. Try again.

### Step 6-2

Moisture within the deck should trigger the dew sensor, which shuts everything down to prevent damage. If the VCR is equipped with a dew light, check to see if it is on. Otherwise, visually inspect the interior of the VCR for signs of moisture. Look particularly at the polished metal surface of the video-head drum. It will show condensation more readily than other parts. If dew is present, leave the VCR on for 30 minutes to one hour. Keeping the VCR powered warms up the mechanism and helps burn off the moisture.

### Step 6-3

A defective reel tire or belt tire can prevent the take-up reel from spinning. This causes tape spilling, which can result in a jam. To prevent considerable damage to the deck and tape, VCRs are out-fitted with a take-up reel sensor (described more fully in chapter 2, "How VCRs work"). If the take-up reel spindle stops, the machine will cease playback after a 5- to 10-second delay. If the problem is mechanical—the reel isn't being driven by the belt or tire—inspect for slipping rubber and replace or clean the faulty part.

### Step 6-4

The take-up reel sensor works by shining a light toward the inside rim of the take-up reel spindle. The light is reflected off strips of plastic or metal foil and directed to a sensor. If the light or sensor is malfunctioning, or the wires leading to the PCB are broken or shorted, the deck will assume the take-up reel has stopped and will place itself in autostop mode. Playback stops after about 5 to 10 seconds. Check the wires for breaks and shorts.

If possible, you should avoid taking the take-up reel spindle off; the height of the spindle is crucial and shouldn't be altered. If you must remove the reel spindle, be sure to reassemble it with all of its parts in the correct order. Failure to replace a washer or other component will result in severe audio and video problems.

If the reel turns and all other functions look normal, inspect the sensor for dirt or oil and clean it (you must remove the take-up reel spindle for this). If necessary, replace the sensor or sensor electronics.

### Step 6-5

An obstruction blocking the video-head drum can cause the drum to stall out. A stalled drum will prevent video playback, and most likely, will prevent the VCR from playing at all (in most VCRs, the head drum provides the timing signal to operate the drive capstan). The head drum also can be inoperative if the scanner motor or electronics are damaged.

To check for a stalled video-head drum, load the tape and press PLAY. The head should immediately start to spin, even as the tape is threading. If it does not, the drum is stalled. Check for obstructions and clear them.

### Step 6-6

Almost all VCRs use a belt or tire to transfer rotational motion from the capstan motor to the capstan. If the belt is old or broken, the capstan will not operate properly, and playback will be impaired. If the capstan does not move at all, immediately stop the VCR. Always avoid a condition where the video heads spin but the tape does not move.

Check the capstan belt or tire and replace or clean it as necessary. While inspecting the capstan, look at the capstan motor. If it is not turning, it could be defective. Because of the delicate nature of the capstan motor, do not use a battery to test operation. Check the wiring to the motor for breaks and shorts. If you have the service manual, refer to it for the location of the capstan motor power terminals. Use your voltmeter to check the voltage at the capstan. It will vary between 5 and 12 V, depending on the speed of playback and the make and model of VCR.

Also check the capstan pinchroller. If the roller is old or has flat patches, the tape won't be pulled through the transport properly. Replace or clean the pinchroller as necessary.

## Flowchart 7. VCR eats tape

Nothing is worse than a VCR with a voracious appetite. When a VCR eats a tape, the tape is usually ruined beyond repair. You can repair some tapes, but the splice can cause dirty video heads—

7. VCR eats tape

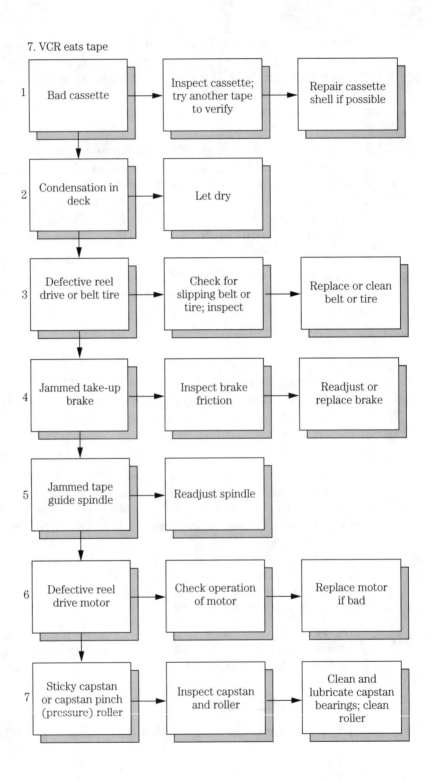

| | | |
|---|---|---|
| 1 | **Bad cassette** | Inspect cassette; try another tape to verify → Repair cassette shell if possible |
| 2 | **Condensation in deck** | Let dry |
| 3 | **Defective reel drive or belt tire** | Check for slipping belt or tire; inspect → Replace or clean belt or tire |
| 4 | **Jammed take-up brake** | Inspect brake friction → Readjust or replace brake |
| 5 | **Jammed tape guide spindle** | Readjust spindle |
| 6 | **Defective reel drive motor** | Check operation of motor → Replace motor if bad |
| 7 | **Sticky capstan or capstan pinch (pressure) roller** | Inspect capstan and roller → Clean and lubricate capstan bearings; clean roller |

274

even video head damage—during subsequent playback. The best way to avoid a VCR that eats your tapes is regular preventive maintenance. But if your VCR is having your tapes for dinner, check the possible cures that follow.

## Step 7-1

Inspect the cassette to make sure it is not defective. The tape might be catching on the cover and is unable to come out of the shell, or the internal reel locks in the shell could be preventing the tape from moving. To test the cassette, open the cover (press the release button on the side). With Beta cassettes, opening the hatch disengages the reel locks, so you can now turn the reels to test the free movement of the tape. With VHS cassettes, you must depress the reel lock button, located in the center on the underside of the shell. Turn the reels toward the center of the cassette. Try both reels to make sure that the tape moves freely back and forth within the cassette.

If the tape won't move, it might be binding inside the shell or the locks are broken. Disassemble the shell, as described in chapter 5, "General cleaning and preventive maintenance."

## Step 7-2

Moisture within the deck can cause the tape to catch on the polished surface of the head drum. Even if the deck is outfitted with a dew sensor, in some instances it will play, but if the tape stops, considerable tape stretching and damage occurs.

If the VCR is equipped with a dew light, check to see if it is on. Otherwise, visually inspect the interior of the VCR for signs of moisture. If dew is present, leave the VCR on for 30 minutes to one hour. Keeping the VCR powered warms up the mechanism and helps burn off the moisture. You also can use a hair dryer set on low heat or no heat.

## Step 7-3

A defective idler tire can prevent the take-up reel from spinning, causes tape spillage because the capstan continues to feed tape to the take-up reel, but the reel is not winding up the tape fast enough or not at all. Replace or clean the take-up reel idler tire. When cleaning, follow the recommended procedure outlined in chapter 5, "General cleaning and preventive maintenance."

## Step 7-4

The supply and take-up spindles both have brakes (sometimes referred to as torque limiters), as shown in Fig. 9-19. In most VCRs during normal recording and playback, the supply spindle brake barely touches the reel. It is controlled by the back-pressure lever. If the pressure lessens, the brake clamps to slow the supply reel; when the pressure increases, the brake releases.

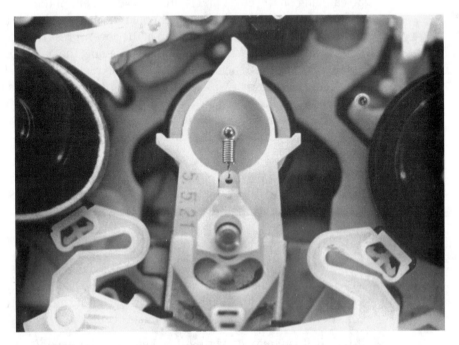

■ **9-19** *Closeup of the tape reels, brakes, and idler wheel pivot.*

Check the supply reel brake to make sure the pad has not worn off and there is slight pressure on the reel during operation. With the cover off, watch the tape as it passes by the back-pressure lever. It should wobble slightly.

Also during normal recording and playback, the brake on the take-up reel is set so only clockwise rotation is possible. If it is grinding against the take-up reel, inspect the mechanism for a jam or look for a loose or broken spring. Free the jam or fix the spring and the brake should operate smoothly.

## Step 7-5

The tape-guide spindles thread the tape around the video heads and other components. If these spindles are broken or bent or out of adjustment, the tape can slip out and become tangled in the works. Inspect the guide spindles for proper operation. Use a spare tape and watch the action of the guides as you thread and unthread the tape. Keep your hand on the power switch and flick it off the moment you see the tape slip off. Manually cycle the threading mechanism back into place and eject the tape. If the guides are broken, replace them. Be careful when adjusting the spindles. They require careful adjustment, as explained in this chapter.

## Step 7-6

A defective reel-drive motor has the same effect as a worn-out reel-drive idler tire. If the take-up reel doesn't turn during playback, tape spillage results. Test the operation of the motor by placing the deck in FAST FORWARD mode. If the take-up reel doesn't move, inspect the idler tire first, then examine the motor.

Note that in most machines, however, the idler tire is driven by the capstan motor. Visually inspect the operation of the capstan motor and examine the belts from the motor to the idler tire. The drive belt is typically on the bottom of the deck, which means you must remove the bottom panel to gain proper access.

If your deck has a separate reel-drive motor, ensure that the proper drive signals are being applied to the motor by connecting a volt-ohmmeter to both motor terminals. Select the dc volts function on the meter, with a range of no less than 12 V. With the deck plugged in and turned on, depress the FAST FORWARD button. With most machines, your meter should read a voltage (it can be as small as 3 V but as high as 12 V).

If voltage is reaching the motor, it might be the motor itself. You can test the motor by connecting a C- or D-size battery to the motor terminals. The polarity of the battery determines which direction the motor turns, so if the motor seems to labor when power is applied in one polarity, stop and try the other direction.

## Step 7-7

A sticky capstan or pinch (pressure) roller also can cause tape spillage. With the top of the VCR off, load a tape and play it. Watch the capstan carefully and look for irregularities. Usually, problems with the capstan show up as video and sound problems, so if you notice these, suspect the capstan or roller.

Clean the capstan and pinchroller. The pinchroller might need to be replaced or rejuvenated. See chapter 5, "General cleaning and preventive maintenance," for information on cleaning and rejuvenating capstan pinchrollers. Removing the roller for replacement or rejuvenation can be difficult because it is mounted on the support stem in different ways. Some VCRs use a screw, as shown in Fig. 9-20, to hold the pressure roller in place. Others use an E clip and a few use friction fit. Analyze the mounting technique before tugging on the roller.

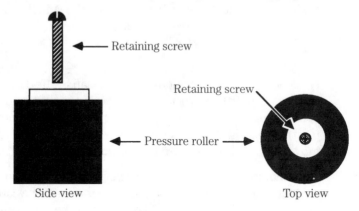

Retaining screw

Retaining screw

Pressure roller

Side view

Top view

■ **9-20** *Assembly detail of the pressure roller.*

Lubricate the capstan bearings by applying oil to the sleeve bearings, as shown in Fig. 9-21. An oiler with a syringe applicator allows you to squirt the lubricant almost directly into the bearings. Use only a small amount of oil.

■ **9-21** *Lubricate the capstan shaft bearings with light machine oil.*

# Flowchart 8. Fast forward or rewind won't operate

The VCR plays and records, but pressing the REWIND and FAST FOR-WARD buttons has no effect. The problem is almost always related to a bad cassette, a dirty or worn-out roller, or a dirty switch.

## Step 8-1

Inspect the cassette to make sure it is working properly. When in doubt, try a known, good tape.

## Step 8-2

A defective idler tire can prevent the take-up or feed reels from spinning, which prevents rewinding or fast forwarding of the tape.

8. Fast forward or rewind won't operate

| | | | |
|---|---|---|---|
| 1 | Bad cassette | Inspect cassette; try another tape to verify | Repair cassette shell if possible |
| 2 | Defective reel drive or belt tire | Check for slipping belt or tire; inspect | Replace or clean belt or tire |
| 3 | FF/rewind switch dirty or broken | Check switch operation with meter | Replace or clean switch |
| 4 | Control mechanism or solenoid jammed or broken | Inspect mechanism for jamming | Clear obstruction or lubricate mechanism |
| 5 | Defective reel drive motor | Check operation of motor | Replace motor if bad |

*Flowchart 8. Fast forward or rewind won't operate*

Replace or clean the idler tire. When cleaning, follow the recommended procedure outlined in chapter 5, "General cleaning and preventive maintenance;" also inspect the outer rim of the supply and take-up reel spindles. Use alcohol to remove any build-up of rubber and grit.

## Step 8-3

The switches on even a solenoid-driven deck are inherently mechanical, and they can get broken, dirty, and corroded. Test the FAST FORWARD and REWIND buttons to make sure they are working properly. Connect the test leads of the ohmmeter to either side of the switch terminals. The meter should read zero ohms when the switch is in the ON position, and infinite ohms when the switch is in the OFF position. If the switch is the double-pole type—two sets of switch contacts inside instead of just one—test each set separately.

If the switch seems okay, trace the wires back to the printed circuit board or solenoid. Use your meter to look for breaks or shorts in the wire.

## Step 8-4

VCRs use two methods of activating the rewind/fast forward mechanics: mechanical levers and electronic control. If your deck is an older model, with "piano" key controls, check the mechanism for jamming and binding. Look for foreign objects that could be blocking the smooth operation of the mechanism. Clean the linkages with a soft, clean rag and lubricate the assembly. Be on the lookout for bent parts and broken or missing springs.

Electronic controls can route to a solenoid, motor, or printed circuit board (or all three, as shown in Fig. 9-22). Check the wires and look for shorts and open circuits. Use your volt-ohmmeter in the usual way. Observe the voltage levels at the solenoids and motor(s) when you press the FAST FORWARD and REWIND keys. If you get a voltage, you know the circuit is working and the wiring is good; it's the solenoid or motor that's broken.

Some VCRs use a solenoid to position the idler tire in REWIND or FAST FORWARD position. If the reel motor turns but nothing happens, check the operation of the solenoid. Is it pulling in? If not, it might be burned out and must be replaced. If the solenoid is attempting to pull in but doesn't, look for an obstruction blocking

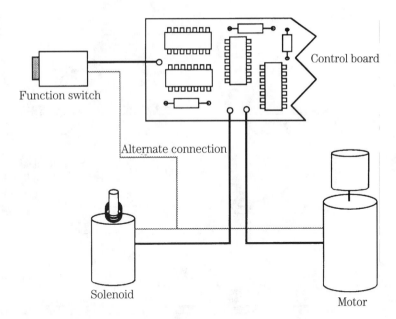

**■ 9-22** *Layout of a function switch and control motor/solenoid.*

the path of the solenoid plunger and mechanics or a broken spring or lever.

**281**

### Step 8-5

In some cases, the reel motor itself will be defective. Check the voltage to the reel motor using your volt-ohmmeter. You should get a reading of between 3 and 12 V. If there is a reading, but the shaft doesn't turn, suspect the motor. On decks that don't have a separate motor for the reel-drive (using a pick-off from the capstan motor), check the belts for dirt, glaze, or breakage.

## Flowchart 9. Search does not work correctly

Most VCRs have a fast scan or search capability that lets you quickly preview the contents of the tape. The search speed varies between models and depends on the speed of the original recording, but it is anywhere from 3 to 15 times faster than normal play. The search function can go on the fritz, however, for any number of reasons. The ones that follow are the most common.

### Step 9-1

Bad cassette. As usual, try another cassette to test your theory.

9. Search does not work correctly

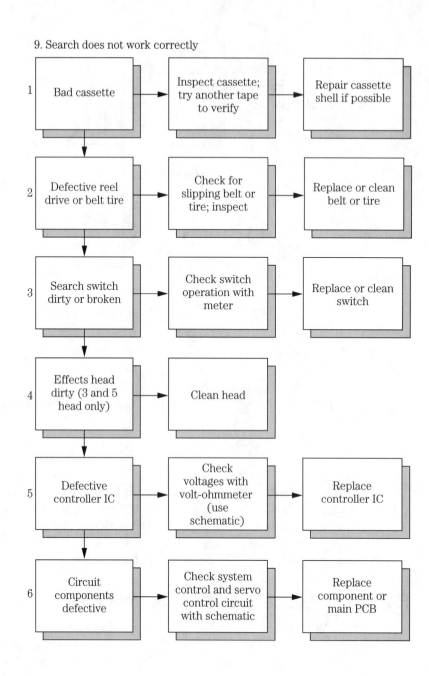

| | | | |
|---|---|---|---|
| 1 | Bad cassette | Inspect cassette; try another tape to verify | Repair cassette shell if possible |
| 2 | Defective reel drive or belt tire | Check for slipping belt or tire; inspect | Replace or clean belt or tire |
| 3 | Search switch dirty or broken | Check switch operation with meter | Replace or clean switch |
| 4 | Effects head dirty (3 and 5 head only) | Clean head | |
| 5 | Defective controller IC | Check voltages with volt-ohmmeter (use schematic) | Replace controller IC |
| 6 | Circuit components defective | Check system control and servo control circuit with schematic | Replace component or main PCB |

## Step 9-2

A defective reel-drive tire can prevent the take-up or feed reel from spinning, and therefore the tape won't shuttle through the VCR. Replace or clean the take-up reel idler tire. Also inspect the capstan pinchroller and clean or replace it, as required.

### Step 9-3

Test the search switches to ensure they are working. Connect the test leads of the ohmmeter to both sides of the switch terminals. The meter should read zero ohms when the switch is in the ON position, and infinite ohms when the switch is in the OFF position.

### Step 9-4

On three- and five-head decks, the extra head is often used for special effects. If this head is dirty, it will appear as if the search function is not working properly. Clean all heads and try again. Some decks use the full complement of four heads to achieve clear fast-scan tape searches, though the two sets of heads are normally used for playback at various speeds. Again, clean all heads and try a known, good tape.

### Step 9-5

A bad controller IC can prevent searching. Check voltages with a volt-ohmmeter. You will need a service manual or schematic to perform this operation. If the voltages aren't what they are supposed to be, replace the controller IC.

### Step 9-6

Additional circuit board components can be defective. As before, check the voltages with a volt-ohmmeter against the specifications in the service manual. Replace the faulty component or circuit board as required.

## Flowchart 10. Tracking control has no effect

Beta and VHS VCRs have a tracking control that helps you lock onto the picture when playing tapes made on other machines (8mm VCRs don't have a tracking control, as explained in chapter 2, because they automatically sense the correct tracking pattern on the tape). The tracking control also is sometimes used to view tapes you made with your VCR but have gotten old or stretched with age. Sometimes, turning the tracking control has no effect. Here are some causes.

### Step 10-1

Is the tape too bad to use? If you can, try the tape on another deck. If it plays well on another VCR but not in yours, you can proceed with the troubleshooting.

10. Tracking control has no effect

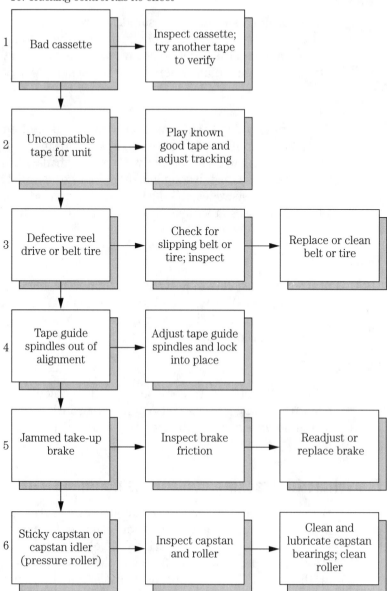

**Step 10-2**

Although you might be using the proper format of cassette for your deck, the tape could have been recorded on an incompatible machine. These include tapes made in England and other countries that use a different broadcast standard than the U.S. The broadcast standard in the U.S. is NTSC (National Television System Committee); many other countries use incompatible broadcasting standards known as PAL (phase alteration line), PAL-M, and SECAM (sequential and memory).

In addition, tapes made on a Super VHS deck and recorded in SU-PER mode will not play on regular VHS VCRs.

### Step 10-3

A defective idler tire can prevent the take-up reel from spinning and thus will impair playback. Replace or clean the idler tire, as recommended in chapter 5.

### Step 10-4

The tape-guide spindles thread the tape around the video heads and other components. If these guides are broken or misaligned, the tracking might be off. To realign the spindles, follow the procedure given in this chapter. Be careful when adjusting the guide spindles because you can easily make things worse if you are careless.

### Step 10-5

The supply and take-up reels both have reel brakes, as shown in Fig. 9-19. With most VCRs, the brake on the supply reel barely touches the reel, providing back pressure during recording and playback. Check this brake to make sure the pad has not worn off and there is slight pressure on the reel. The brake on the take-up reel is disengaged during playback and recording and allows only clockwise rotation.

**285**

### Step 10-6

A sticky capstan and pinchroller also can cause mistracking as well as profound audio and video troubles. Watch the capstan carefully and look for irregularities. Clean the capstan and pressure roller with alcohol. If the pressure roller needs rejuvenation, refer to Flowchart 7 for details. Additionally, oil the capstan bearings with a syringe applicator.

## Flowchart 11. VCR does not respond to some or all front-panel controls

Failure to respond to some or all of the front-panel controls is often a mechanical fault, but it can also be caused by the electronics on the main PCB. If not all of the switches are affected, the problem might lie in the switches themselves (or the functions associated with that switch). If none of the switches respond, refer to Flowcharts 1, 2, and 6.

11. VCR does not respond to some or all front-panel controls

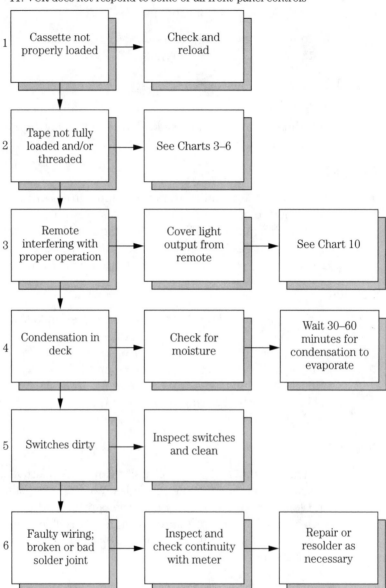

## Step 11-1

If the tape is not properly loaded, the front-panel controls might not appear to work properly. Remove the tape and reinsert it. A damaged tape can also cause the controls to behave erratically.

286

## Step 11-2

In most VCRs, the tape must be completely loaded and threaded before the front-panel controls operate. Check Flowcharts 3 through 6 if the tape is not loading and threading properly. Be sure to check the cassette-in switch (usually a small leaf switch located on the inside of the cassette-lift mechanism). The switch could be dirty or broken.

## Step 11-3

If your VCR has a remote control, it might be overriding the front switches. To rule out the possibility of a faulty remote control, remove its batteries, take it to another room, or—if the control is the wireless infrared type—cover up the end of it to block the light. If the control is the wired type, disconnect it from the deck. Refer to Flowchart 18 if disabling the remote frees the operation of the front-panel controls on the deck (it is an indication that the remote is faulty).

## Step 11-4

Taking a VCR from the cold, damp outside air to the warm, inside air can cause condensation to form on the internal components. Most VCRs (especially portable units and camcorders) have a dew sensor that prevents the machine from operating when condensation is present. If you suspect condensation has formed inside the machine, visually inspect for moisture and wait 30 to 60 minutes with the power on for it to evaporate. You can speed up the drying process somewhat by using a hair dryer, but keep the dryer on low or no heat.

## Step 11-5

Given the right set of circumstances, switches can get broken, dirty, and corroded. Test each switch on the front panel to make sure all work properly. Connect the test leads of the ohmmeter to either side of the switch terminals. As you depress the switch, the meter should alternately read zero or infinite ohms. If not, it is an indication that the switch is faulty.

Unsealed switches can be cleaned using an electrical contact spray cleaner. Liberally squirt the cleaner inside the switch. Broken switches must be replaced.

## Step 11-6

If the switches themselves check out, the problem might be caused by faulty wiring. Carefully inspect the wiring leading to and from the front-panel control switches. Look for obvious breaks, kinks, and cold solder joints. Use a meter to test the continuity of all affected switches.

The wiring harness connecting the front panel PCB and the main PCB might be the flexible membrane type. The connecting wires are copper bands that are glued onto a flexible mylar base. These flexible ribbon connectors can be easily damaged if abused and are extremely difficult to repair. If you suspect a broken trace on the ribbon, double check the continuity using a volt-ohmmeter. A faulty ribbon must be replaced.

## Flowchart 12. Front-panel indicators not functioning

You put in a cassette, press PLAY, and everything works. That is, everything but the front-panel indicators. Check the following points if the indicators in your VCR don't light.

12. Front-panel indicators not functioning

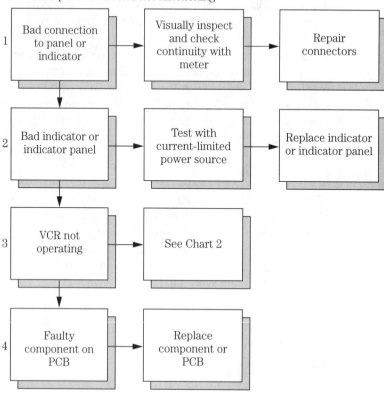

288

## Step 12-1

The indicators used in most VCRs are the LED (light-emitting diode), LCD (liquid-crystal display), or fluorescent type. These have exceptionally long life spans, so it is unlikely that they have simply "burned out" like light bulbs do. The most probable cause is a bad wire or connection leading to the front panel or indicator.

Visually inspect all the wires leading to the front-panel PCB and the indicator modules themselves. Use a volt-ohmmeter to test for continuity. With the test leads connected at the ends of each wire, the reading should be 0 ohms. A reading of infinite ohms indicates that a wire is broken or that one of the solder joints is bad. Replace the wire or resolder as necessary.

The wiring harness connecting the indicators and the front-panel or main PCB might be the flexible membrane type. The connecting wires are actually copper bands glued onto a flexible base. These flexible ribbon connectors can easily be damaged and are difficult to repair. If you suspect a broken trace on the ribbon, double check the continuity using a volt-ohmmeter. Replace if bad.

## Step 12-2

Rarely will all the LEDs and number segments go out all at once. The more common occurrence is that only one indicator lamp or one segment in one numeral will fail. Check first to make sure that the wiring is not at fault. Use the volt-ohmmeter as before. You also can test LED and LCD indicators and segments using the current-limited power source shown in Fig. 9-23. Don't forget the resistor; connecting the battery directly to the connections on the panel will burn out the segments.

If one or more indicators or segments are indeed bad, the entire module must be replaced. You cannot repair or replace individual indicators or segments within the module.

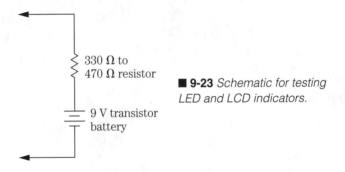

330 Ω to
470 Ω resistor

■ **9-23** *Schematic for testing LED and LCD indicators.*

9 V transistor
battery

### Step 12-3

A blank indicator panel might also be due to a fault on the deck. Be sure that the VCR responds properly to all the front-panel controls. If it does not, refer to Flowchart 11 for more information.

### Step 12-4

If all checks out so far, the fault might be a bad component on the front panel or main PCB. If the VCR works fine in every other regard but the indicator panel is blank or scrambled, the problem might be faulty driver ICs or other components on the front panel or main PCB. Should this be the case, the components or the PCB must be replaced.

## Flowchart 13. Timer does not operate properly

Almost all VCRs, except the very first models, incorporate a timer function where you can command the deck to record at a specific time of day (and often a specific channel). The timer feature is controlled either by the microprocessor or by discrete logic components. In most cases, programming problems are caused by mechanical defects and can be readily repaired.

### Step 13-1

As always, check to make sure that the tape is properly loaded and that the loading mechanism is working. Test the deck for proper operation by pressing the PLAY button and watch the picture (obviously, use a previously recorded tape for this, not a blank one). If the tape plays, you know that everything is fine to this point.

You can try the timer by entering start and stop times of just a minute or two. Set the timer to go off in a few minutes and wait patiently. Watch the clock on the front of the VCR and note if the machine is actuated at the appropriate time. (All this assumes you know how to use the timer function. The timer is one of the most difficult functions to use in any VCR, and you might need to carefully read the manual to make sure that you are doing everything properly.)

### Step 13-2

VCRs sometimes ignore timer instructions if the timer is already set to record something. Turn the timer off, and then enter new instructions.

13. Timer does not operate correctly

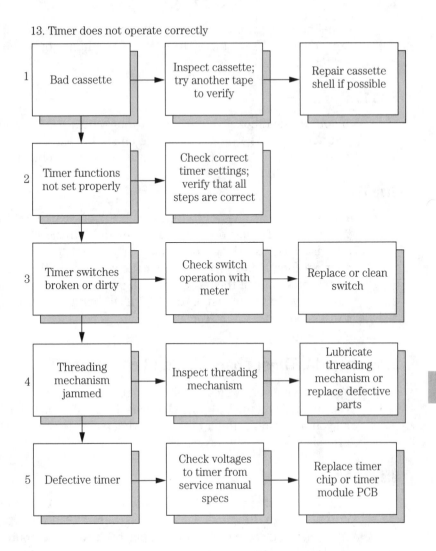

## Step 13-3

The switch(es) that control the timer functions might be broken, dirty, or corroded. Test all the switches to make sure they work properly. Connect the test leads of the ohmmeter to either side of the switch terminals. When the switch is depressed, the meter should alternate between zero and infinite ohms. If it does not, there is a good chance that the switch under test is defective. If the switch is unsealed, you can clean its contacts with an electronic contact cleaning spray.

If the switches check out, the problem might be caused by faulty wiring. Carefully inspect the wiring leading to and from the front-panel control switches. Look for obvious breaks, kinks, and cold solder joints. Use a meter to test the continuity of all affected switches.

### Step 13-4

If you set the timer, go away, and on your return you find the VCR did not record the program, double check the threading mechanism. See Flowchart 5 if you suspect a threading problem.

### Step 13-5

Finally, if the switches and wiring check out, it's safe to assume the problem lies further on in the timer module. If the clock to the VCR is also not working, you can be fairly sure that the timer is either not functioning properly or is not connected to the main PCB or power supply of the deck. Trace the wiring to be sure that the timer is wired to the rest of the VCR. If the wiring seems okay, the timer module should be replaced.

## Flowchart 14. Tuner channels don't change on TV

You are using your VCR as a television tuner. But when you change the channels on the VCR, the channels on the TV don't change, or all you see is static. Refer to the following points for recommended remedies.

### Step 14-1

Check the position of the TV/VCR switch. When in TV position, the output of the VCR's tuner is connected to the TV. In VCR position, the tuner is disconnected (during playback) and the contents of the tape is shown.

### Step 14-2

The VCR must be on for the VCR tuner to work. Be sure there is power to the deck.

### Step 14-3

Many VCRs let you change the output channel between channels 3 and 4 (some even let you choose other channels, such as 5 and 6). Be sure the position of the switch agrees with the channel you are tuned to on the TV. In most instances, the switch and TV channel are set to channel 3.

14. Tuner channels don't change on TV

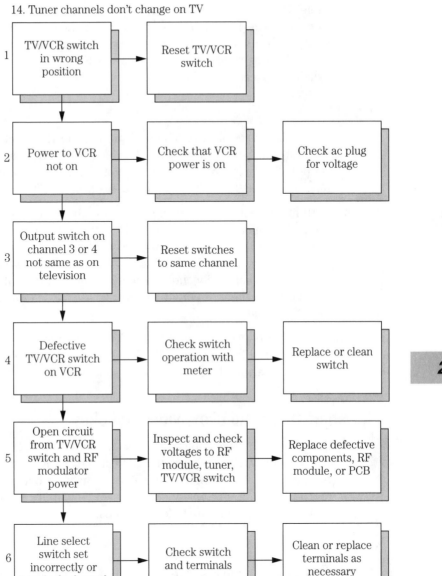

## Step 14-4

The TV/VCR switch could be bad. Check the switch with a volt-ohmmeter.

## Step 14-5

The RF modulator or tuner sections of the VCR could be defective. First, isolate the problem by connecting the TV to the antenna or cable. Play a tape to make sure that the TV is receiving signals from the VCR. If these items check out, inspect the RF terminals on the TV for damage, shorts, and open circuits. Defective components must be replaced; the tuner and RF modulator sections are usually replaced as a whole.

### Step 14-6

Almost all VCRs have two sets of inputs: the RF inputs and the VIDEO IN and AUDIO IN inputs. The deck can handle signals presented at only one of these inputs, not both. The deck will either have a LINE SELECT switch or will have automatic line switching.

With line select, you push the button to choose between the RF input or the AUDIO/VIDEO input. Check to make sure that the switch is set to the proper position, depending on how you have your VCR wired up. On machines with auto line switching, the deck will automatically choose the AUDIO/VIDEO inputs if a cable is connected to them. If you don't want to use these inputs, disconnect the cables.

If the AUDIO/VIDEO terminals on the VCR are shorted, the VCR will disengage the RF input and show only snow or a strange picture. Check the terminals to determine if shorting has occurred.

## Flowchart 15. Sound okay; video not okay

The sound plays back, but the video is either nonexistent, washed out, or snowy. Bad video is usually caused by dirty video heads, but you also should check the cable connection between the TV and VCR. Other circumstances can also contribute to poor video and are explained in the following steps.

### Step 15-1

Check to see if the cassette is bad. Try a known good tape.

### Step 15-2

Other than a bad tape, dirty video heads are the leading cause of video problems. Follow the procedure given in chapter 5, "General cleaning and preventive maintenance," on how to correctly clean video heads. Cleaning video heads improperly can permanently ruin them, so be sure you know what you are doing before undertaking the task.

15. Sound okay; video not okay

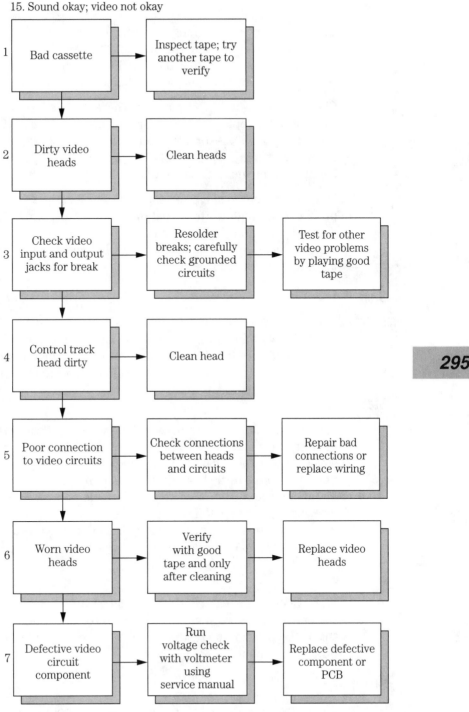

**1** Bad cassette → Inspect tape; try another tape to verify

**2** Dirty video heads → Clean heads

**3** Check video input and output jacks for break → Resolder breaks; carefully check grounded circuits → Test for other video problems by playing good tape

**4** Control track head dirty → Clean head

**295**

**5** Poor connection to video circuits → Check connections between heads and circuits → Repair bad connections or replace wiring

**6** Worn video heads → Verify with good tape and only after cleaning → Replace video heads

**7** Defective video circuit component → Run voltage check with voltmeter using service manual → Replace defective component or PCB

### Step 15-3

Check the cables stretching between the VCR and TV. If you are using the RF terminals on the deck and TV, be sure the coaxial cable is not crimped and that the F fittings are on tight. Inspect the center conductor on both ends; it should be bright and shiny.

Check the cable with your meter. There should be no shorts or open circuits. Replace the cable or remake the F fittings if they are bad (see appendix D) or if the center conductor is dirty or broken. If you are using the direct VIDEO output of the VCR, check the cable in the same manner as above.

Dirty or broken terminals on the VCR or TV can also cause video problems. Enough of the signal might be getting through to pass the audio portion, but the connection isn't good enough for the picture. Clean the terminals using alcohol and a cotton swab and inspect them for damage.

### Step 15-4

A dirty control-track head can cause the video to look garbled. Clean the head and try again. In addition, try adjusting the tracking control on the front panel of the VCR to clear the lines in the screen. If the tracking control has little or no effect, refer to Flowchart 10.

### Step 15-5

A broken or shorted wire between the video heads and the video circuitry can cause weak or no video output. The wires are usually small, so test them carefully. In many VCRs, a video amplifier is housed on or near the head drum. Check the wires from the heads to the amp and from the amp to the video-processing circuitry elsewhere on the VCR. Look for bad connections (clean the connectors and make sure they are tight), broken wires, and shorts. Remake any bad solder joints.

### Step 15-6

Worn video heads can cause an overall washed-out look. Because the only cure for worn video heads is to replace them, make sure this is the problem by testing many tapes and troubleshooting all other components.

### Step 15-7

A defective component in the video preamplifier or video-processing circuits can produce weak or no video. To determine if a component is bad, run a voltage check with a voltmeter. Use a service manual or schematic to trace the circuit accurately. Defective parts can be replaced; otherwise, exchange the entire printed circuit board.

## Flowchart 16. Video okay; sound not okay

The VCR does everything it should do with one important exception: there is no audio. As you will see in this section, most audio dysfunctions are caused by dirty heads or poor cabling.

### Step 16-1

A stretched or damaged tape can cause the sound to waver in and out or flutter annoyingly. In some extreme cases, the tape is so

16. Video okay; sound not okay

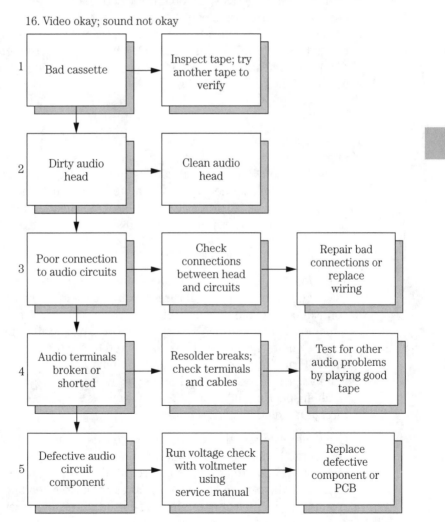

damaged that no audio can be heard. Check the operation of the VCR by using a known, good tape. If the problem persists, then the reason lies in the VCR and not the tape.

### Step 16-2

Dirty audio heads are a major cause of audio problems. Thoroughly clean the heads as described in chapter 5, "General cleaning and preventive maintenance." Linear stereo VCRs have a single audio head unit with many elements built into it (including the control-track head and both channels of audio). Remember to clean the entire surface of the head unit.

VHS hi-fi decks use two separate rotating heads to deliver hi-fi sound. Clean these as you would the video heads. Refer to chapter 5 on how to properly clean the rotating audio heads; they are as fragile as the video heads and can be seriously damaged if mistreated. Beta hi-fi decks use the video heads to record stereo audio. If these heads are clogged, odds are that both the picture and sound will be bad.

### Step 16-3

Loose audio cables between the VCR and the TV or hi-fi is a common cause of no sound. Visually inspect the cables and test them using a meter. Replace or repair bad cables. If your VCR has a headphone jack, connect a pair of phones to it. You'll hear sound if the deck is operating properly. On decks with a headphone volume control, be sure to turn the control up so you can hear the audio.

### Step 16-4

The audio input or output terminals might be broken or shorted. If the terminals or the internal wires leading to them are shorted out, you will not be able to hear sound. However, you might hear a low humming sound through the speakers or TV. The humming can occur even if your hi-fi or TV is connected to the RF output jack of the VCR and not the direct audio terminals. If this occurs, immediately turn the deck off and unplug it. Remove the cover and visually inspect in and around the audio jacks for shorted wires or foreign metallic objects.

Use a meter to test the resistance of the jacks. With the test leads connected to the inner and outer connectors of one jack, the meter should measure at least some resistance (on some decks and terminals, the reading could be near zero ohms). A reading of 0 $\Omega$ is an indication that the terminal is shorted. If the cause cannot be found, the audio output circuit must be serviced. With some decks, the circuit is on its own printed circuit board; in others, the audio output circuitry is contained on the main PCB.

Loose, shorted, or grounded audio cables between the VCR and the monitor is a common cause of hum and distortion. Visually inspect the cables and test them using a meter. Replace or repair bad cables. If your VCR has a headphone jack, connect a pair of headphones to it. The sound should be clear and undistorted.

Excessive hum also can be caused by poorly shielded signal cables or cables routed close to ac power cords. The hum is caused by the close proximity of the alternating-current field. If hum is a problem, examine the quality of the cables. The better the cable, the better the shielding, and the greater the rejection of ac-induced hum. Route the cables as far from ac cords as possible, and keep the cable lengths as short as possible.

Dirty phono jacks on the VCR or monitor can impede the signal and cause noticeable audio distortion. Clean the jacks with an electronic contact cleaner. Inspect the cable connectors, and clean them if they look dirty.

### Step 16-5

Use your meter to check the wiring between the audio head(s) and the circuits on the PCB. If the wiring passes scrutiny, refer to a service manual or schematic and use the volt-ohmmeter to check for proper voltages on the circuit boards. Some decks have provisions for altering the audio input and output levels; change these controls only when you have access to the service manual data. Replace components that appear faulty or replace the entire PCB.

## Flowchart 17. Snowy video; poor audio

When troubleshooting snowy video and scratchy audio symptoms, be wary of problems in the tape, TV, and connections. It is in these areas where the problems usually reside.

### Step 17-1

A stretched or damaged tape can cause serious sound and/or picture disabilities. The cassette shell also might be at fault, restricting the free movement of tape and causing excessive stress. Check the operation of the VCR by using a known good tape (a prerecorded movie is a convenient and functional choice). If the problem persists, then the reason is with the VCR, not the tape.

17. Snowy video; poor audio

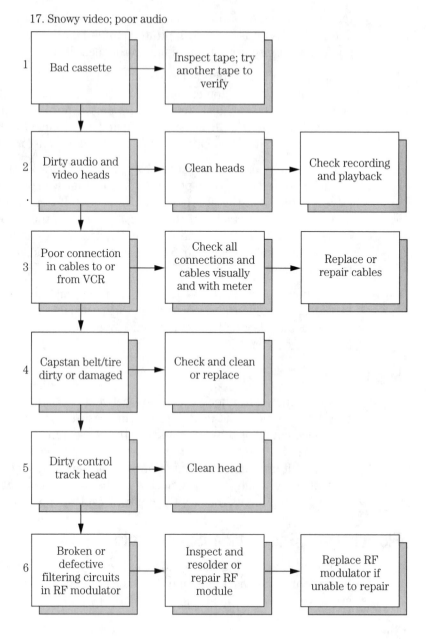

| | | | |
|---|---|---|---|
| 1 | Bad cassette | Inspect tape; try another tape to verify | |
| 2 | Dirty audio and video heads | Clean heads | Check recording and playback |
| 3 | Poor connection in cables to or from VCR | Check all connections and cables visually and with meter | Replace or repair cables |
| 4 | Capstan belt/tire dirty or damaged | Check and clean or replace | |
| 5 | Dirty control track head | Clean head | |
| 6 | Broken or defective filtering circuits in RF modulator | Inspect and resolder or repair RF module | Replace RF modulator if unable to repair |

## Step 17-2

In addition to bad tapes, dirty audio and video heads are a major cause of sound and picture problems. Thoroughly clean the audio and video heads as described in chapter 5, "General cleaning and preventive maintenance." Be sure to get all of the heads. VHS hi-fi decks use two separate rotating heads to deliver hi-fi sound. Clean

300

these as you would the video heads. Be sure to follow the directions given in Chapter 5 on how to clean rotating audio and video heads properly. They can be seriously damaged if mistreated.

## Step 17-3

Unsecured audio cables between the VCR and the TV or hi-fi is a common cause of poor sound and picture. Visually inspect the cables and test them using a meter. Replace or repair bad cables. If your VCR has a headphone jack, connect a pair of headphones to it. You'll hear sound if the deck is operating properly.

## Step 17-4

VCRs typically use a belt or tire to drive the capstan from the capstan motor. If the rubber of the belt or tire is worn or broken, the capstan will not operate properly and playback will be impaired. Check the capstan belt or tire and replace or clean it as necessary. While inspecting the capstan, look at the capstan motor. If it is not turning, it could be defective. Because of the delicate nature of the capstan motor, do not use a battery to test operation. Check the wiring to the motor for breaks and shorts.

A sticky capstan or pinchroller also can cause playback problems. Load a tape, play it, and watch the capstan carefully and look for irregularities. Clean the capstan and pressure roller. The pressure roller might need to be replaced or rejuvenated. See chapter 5, "General cleaning and preventive maintenance," for information on cleaning and rejuvenating capstan pinchrollers.

Removing the roller for replacement or rejuvenation can be difficult because they are mounted on the support stem in a variety of ways. Some VCRs use a screw to hold the pressure roller in place. Others use an E clip and a few use friction fit. Analyze the mounting technique before tugging on the roller.

Lubricate the capstan bearings by conservatively applying oil to the sleeve bearings. An oiler with a syringe-type applicator works best and helps you apply the lubricant deep into the bearings.

## Step 17-5

Dirty control-track heads can cause the tape to slow down and speed up or even travel through the VCR at wild speeds. Clean the control-track head thoroughly and try again. Adjust the tracking control to eliminate or reduce the lines in the picture. If the tracking control can't be adjusted, refer to Flowchart 10.

*Flowchart 17. Snowy video; poor audio*

### Step 17-6

Broken or defective filtering circuits in the RF modulator can cause sound and picture problems. If your TV has direct audio and video inputs, try connecting the VCR to them. If the problem goes away, the RF modulator needs to be serviced. Inspect the modulator and resolder or repair it as necessary. You must replace the RF modulator if you are unable to repair it.

## Flowchart 18. Remote control does not operate or function properly

Problems with the remote control functions can be due either to a faulty remote transmitter or bad receiver circuits in the deck. Fortunately, problems with the remote control are almost always caused by the transmitter, which is not only easier to repair but cheaper to replace should it be seriously defective.

If the remote control transmitter is the infrared type, you can test for proper operation by using the light sensor described in chapter 4, "Tools and supplies for VCR maintenance." If the controller is the wired type, check it as you would any switch and wire in the VCR. Use a volt-ohmmeter to check for open wires or shorts. See Step 18-6.

### Step 18-1

If the remote control is not working properly, first make sure you are operating the transmitter within the prescribed limits. Most transmitters don't work when operated at distances beyond 20 feet from the deck or at angles greater than 30 degrees off to either side of the receiving sensor. If you must place the deck off-angle to the direct beam of the transmitter, you can try "bouncing" the infrared light off a white card. Place the card in front of the VCR, as shown in Fig. 9-24.

You also can use an infrared repeater, which rebroadcasts the codes from the control. Position the repeater in a central location that is in-line with the controller and VCR.

### Step 18-2

Batteries that are low or dead will obviously cause the controller to fail. Even batteries that have some kick left can cause some trouble, so even after testing them with a meter, try a fresh set to see if it makes a difference.

18. Remote control does not operate or function properly

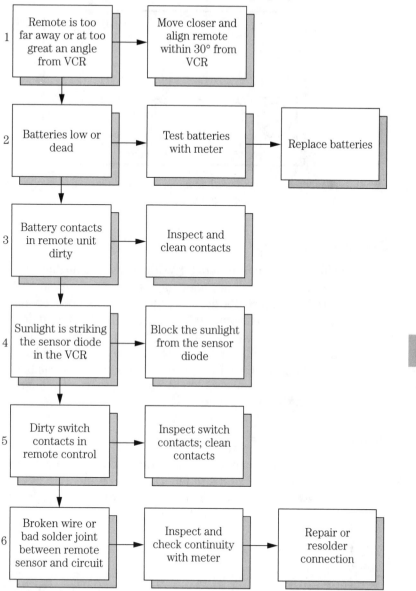

**303**

## Step 18-3

While the batteries are out of the transmitter, inspect the battery contacts. They should be bright and shiny. If not, there might be insufficient current getting to the transmitter circuits. Clean the contacts and reinsert the batteries.

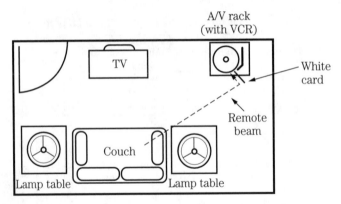

■ **9-24** *Use a white card as a reflector when the VCR is too far off the axis from the normal use of the remote controller.*

### Step 18-4

Most remote controllers for VCRs work by emitting short pulses of infrared light (some older ones use ultrasonic sound). Sunlight striking against the deck might overload the receiving sensor so the commands from the transmitter are not adequately received. Therefore, always shade the sensor from direct sunlight. If the deck is subjected to sunlight, the remote control function still might not operate properly even after the light has been blocked. Wait 5 to 10 minutes for the sensor and surrounding components in the VCR to cool down.

### Step 18-5

Dirty switch contacts in the remote control might cause all or some of the functions to fail. Open the remote and liberally spray the switches with an electrical contact cleaner. Inspect the wiring and solder joints and make repairs as necessary.

### Step 18-6

Broken wires and bad solder joints between the receiver sensor and the main PCB can also cause problems. Open the VCR only after you are certain that the transmitter is operating properly (use the light sensor described in chapter 4 to be absolutely sure). Inspect the wires and solder joints; test the continuity with a volt-ohmmeter. With the test probes connected to either end of each wire, the reading should be zero ohms. A reading of infinite ohms is an indication of an open circuit.

# Flowchart 19. Unusual mechanical noises during loading and playback

Mechanical noises are those terrible grinding, crunching, rattling, and squeaking noises that can emanate from the VCR. A well-designed VCR is nearly silent in operation, and any noticeable noise is indicative of a malfunction inside the unit.

## Step 19-1

Check that the tape is loaded properly.

## Step 19-2

Is the tape damaged? Eject the tape and inspect for warping of the shell or damage to the tape itself.

## Step 19-3

Mechanical defects in belts and pulleys can cause squeaks and other unsavory sounds. Belts and pulleys are often used in the capstan drive, loading, and threading mechanisms. With the top and bottom covers of the deck off and a tape loaded, press the PLAY button and watch the drive motor, pulley, belt, and spindle. You should be able to see any mechanical defects, if any are present.

If a belt or roller seems like it is slipping, clean it (refer to chapter 5 for details), or replace it with a new one. Do the same when loading and unloading a tape. Watch the cassette-lift motor or drive belt and look for signs of trouble.

## Step 19-4

A number of machines use plastic gears to drive the cassette-lift mechanism from a motor. A gear is mounted on the motor shaft and meshes with another gear or a straight-tooth rail. When the motor shaft turns, the gear inches the cassette-lift mechanism along the gear or rail, thus loading or unloading the cassette. Inspect the gearing to be sure that none of the teeth are broken and that the gear teeth are meshing properly.

Even if the cassette-lift mechanism looks fine, the gears and rails can still bind against one another and cause grinding sounds. If necessary, apply a small dab of grease on the gear surfaces or oil the shafts and levers.

19. VCR makes unusual mechanical noises
during loading and playback

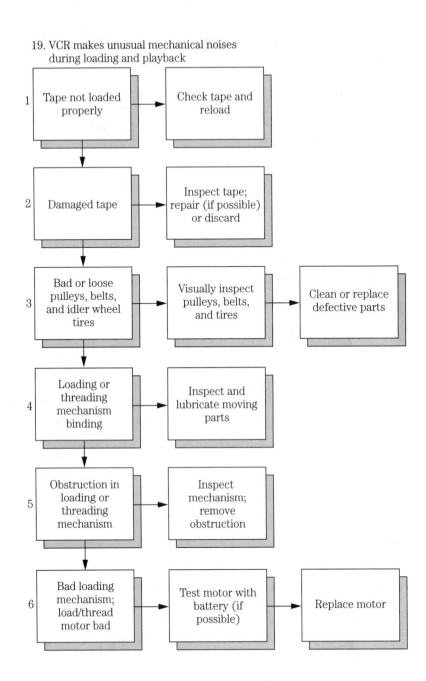

| | | | |
|---|---|---|---|
| 1 | Tape not loaded properly | → Check tape and reload | |
| 2 | Damaged tape | → Inspect tape; repair (if possible) or discard | |
| 3 | Bad or loose pulleys, belts, and idler wheel tires | → Visually inspect pulleys, belts, and tires | → Clean or replace defective parts |
| 4 | Loading or threading mechanism binding | → Inspect and lubricate moving parts | |
| 5 | Obstruction in loading or threading mechanism | → Inspect mechanism; remove obstruction | |
| 6 | Bad loading mechanism; load/thread motor bad | → Test motor with battery (if possible) | → Replace motor |

## Step 19-5

Obstructions in the tape loading or threading mechanisms also
can cause grinding or scratching sounds. Inspect the moving
parts for foreign objects and remove them. Any loose parts should
be retightened.

**Step 19-6**

The motors used to drive the loading and threading mechanisms might be bad. A squeaky or raspy sound is caused by worn bearings. You can isolate the problem to the motor by disconnecting the drive and pressing the PLAY button. Check for any movement in the shaft that could indicate worn bearings. If the motor makes excessive noises while spinning, it's a good indication that it is bad.

# Flowchart 20. You receive an electrical shock when you touch the VCR

The words *thrilling* and *breathtaking* are often used to describe the experience of watching a good movie on a VCR. The word *shocking* should not be in a VCR user's vocabulary unless there is something wrong with the deck.

Do not use a VCR that gives you a shock—even a mild one. Some or all the ac power has been shorted to the metal cabinet, so when you touch the machine, you get a shock (and if the jolt is big enough, you do the "kilowatt dance," which looks a little like the jitterbug but can really put you in the hospital). Before watching another tape, unplug the machine, isolate the cause, and fix it.

You can test for ac leakage current with following procedures. The ac leakage current test determines if any part of the ac line has come in contact with the metal cabinet or base. It is a safety check to prevent a potential shock hazard. It requires the use of a volt-ohmmeter.

With the VCR unplugged, short the two flat prongs on the end of the ac cord, as shown in Fig. 9-25. On the meter, select the 1 k$\Omega$ function and a range of no less than 1000 k$\Omega$. Connect one test lead from the meter to the jumper on the ac cord. Connect the other test lead to any bare (not painted) metal parts of the deck. For a typical VCR, the meter should read about 500 to 1000 k$\Omega$— if you get any reading at all.

A reading substantially lower than 500 to 1000k$\Omega$ is a good indication that the power supply has come into contact with the metal parts of the deck. If this happens, inspect the power cord as it leads into the machine as well as the power switch, the transformer (usually bolted onto the back of the deck), and the wires leading from the transformer to the printed circuit board. If the wires are broken or are shorted to the cabinet, repair the fault before using the deck.

20. You receive an electrical shock
when you touch the VCR

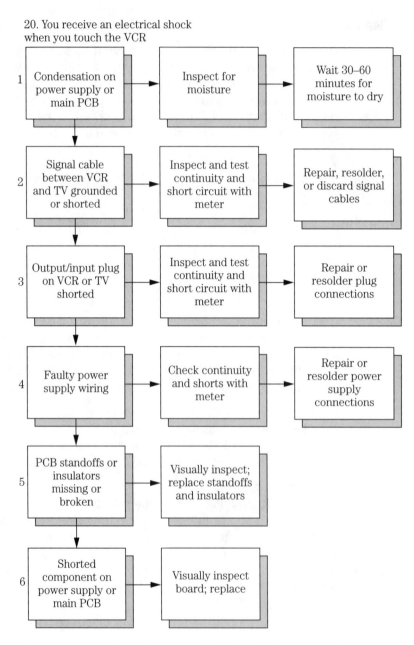

| | | | |
|---|---|---|---|
| 1 | Condensation on power supply or main PCB | Inspect for moisture | Wait 30–60 minutes for moisture to dry |
| 2 | Signal cable between VCR and TV grounded or shorted | Inspect and test continuity and short circuit with meter | Repair, resolder, or discard signal cables |
| 3 | Output/input plug on VCR or TV shorted | Inspect and test continuity and short circuit with meter | Repair or resolder plug connections |
| 4 | Faulty power supply wiring | Check continuity and shorts with meter | Repair or resolder power supply connections |
| 5 | PCB standoffs or insulators missing or broken | Visually inspect; replace standoffs and insulators | |
| 6 | Shorted component on power supply or main PCB | Visually inspect board; replace | |

Not all exposed metal parts of the VCR will return a high value. Touching the center conductor of one of the input or output connectors could yield a lower resistance of 40 or 50 kΩ. This resistance is considered normal with some machines.

An alternative method for checking leakage current is provided in chapter 6, "VCR first aid."

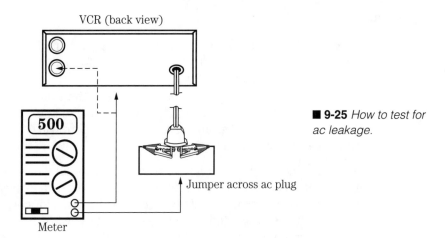

VCR (back view)

■ **9-25** *How to test for ac leakage.*

Jumper across ac plug

Meter

## Step 20-1

Water is a poor conductor of electricity, but it's good enough to short the ac power cord or the power supply terminals against the VCR cabinet. Condensation also can form a light film of moisture that can conduct some current, and although the shock might be slight to the human body, it could damage the VCR electronics.

Inspect for moisture, and if found, let the deck dry for at least 30 to 60 minutes with the power on. You can hurry along the process by using a common hair dryer. Place the dryer in low or no heat only. If there is a lot of water, blot up the excess and spray the interior of the machine with a non-water-based cleaner (see chapters 4, 5, and 6 for suggestions), which has a tendency to displace water and remove it from hard-to-reach places.

## Step 20-2

The signal cable(s) between the VCR and TV might be shorted. Inspect the cabling for obvious damage and test it with a volt-ohmmeter. Connect the test leads across the center and middle conductor of each cable. You should get a reading of infinite ohms. If you get any other reading, it is an indication that the cable is shorted.

## Step 20-3

Determining if the input or output terminals of the VCR is shorted is a little tougher. The resistance values can vary, but most are within a range of about 1 to 50 k$\Omega$ when the test leads are attached across the center and outer connectors of each terminal. A reading of 0 $\Omega$ indicates a possible short.

*Flowchart 20. You receive an electrical shock when you touch the VCR*

### Step 20-4

The wiring leading to and from the power supply could be faulty. Inspect the ac cord leading to the power transformer and power switch. Use your meter to check for short circuits. Examine the power transformer for obvious damage. Check the solder terminals to make sure that none are touching the cabinet. Look closely because even one small strand of wire in the ac cord can cause at least a partial short circuit.

Next, examine the wiring from the transformer to the power supply circuitry. This circuit is sometimes located on its own PCB and other times on the main PCB. Use your meter to check for obvious short circuits. Bear in mind that at least one of the wires from the power transformer will act as the circuit ground for the VCR. You should receive a low reading when checking this wire. The other wires should yield reasonably high resistances.

### Step 20-5

The power supply and main PCBs are generally insulated from the cabinet and base using plastic standoffs. If secured with metal hardware, the boards are electrically isolated from the cabinet with plastic or rubber insulators. Inspect the mounting hardware to see if any parts of the boards are touching the base or cabinet. Inspect the insulators and look for breaks and cracks. Replace any broken standoffs or insulators.

### Step 20-6

Finally, a faulty component on any of the printed circuit boards in the deck can cause some current leakage. Not all components are crucial to the operation of the deck, so even if a part shorts out, the unit might still operate—at least on a marginal level. Visually inspect the components on the boards for obvious damage, and either replace the parts (if possible) or replace the PCB.

## Flowchart 21. VCR overheats

A VCR that gets abnormally hot is not only a potential fire hazard but will exhibit erratic behavior. You can easily tell if the deck is getting warmer than normal by touching its cabinet. It should not be noticeably hot to the touch. If it is, you should investigate the problem before playing any more tapes. Continued use of the VCR might damage it beyond repair.

21. VCR overheats

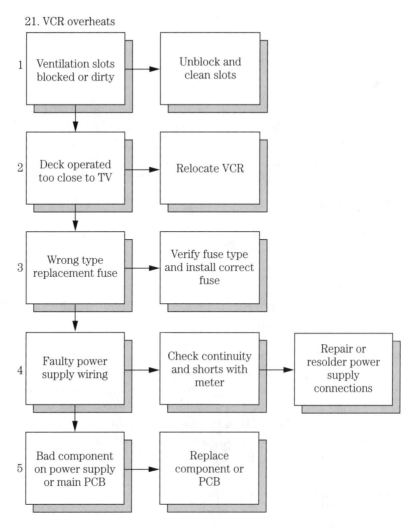

**Step 21-1**

VCRs don't consume much power, so cooling is not as crucial as it is with, say, a 100 W (watt) power amplifier. Still, most decks rely on ventilation slots for proper operation. It is important that these slots not be blocked, either by some outside object or by dust and dirt. If the slots become blocked with dust, use a brush or vacuum to clean them.

**Step 21-2**

A VCR operated too close to the TV might get overly warm because of the heat put out by the set. For proper operation of your VCR, as well as the other components in your video system, you should keep your TV separate as much as possible. Invest in a piece of video furniture (see chapter 3, "The VCR environment") to avoid stacking equipment on top of one another.

### Step 21-3

Almost all VCRs use external or internal fuses. If a fuse blows, it is important to replace it with the same value of fuse as the original. If a higher value is used, the fuse might not blow even though a component in the deck is shorted and drawing excessive current. That can cause fire and considerable damage to the machine.

### Step 21-4

Faulty wiring in the power supply can likewise cause overheating. Check the wiring with a volt-ohmmeter against short circuits.

### Step 21-5

Faulty components on the power supply board or main PCB can cause excessive overheating. You can often identify the responsible component by carefully touching each one. To test, power the deck for a while until the temperature rises, then turn it off. Unplug it from the ac wall socket. Lightly touch each component. None except the voltage regulators should be hot to the touch (some ICs might be warm, but they should not burn your fingers). The voltage regulators, usually mounted on the power supply board, can get hot but are usually kept within a safe operating temperature by the use of aluminum heat sinks.

## Miscellaneous VCR difficulties

Not every VCR malfunction can be neatly categorized. Here are a number of problems that can occur and how to remedy them.

### Will not record

If the VCR operates well in all other respects but does not record, first check the *record tab* on the back of the cassette. If the tab is out, as shown in Fig. 9-26, the record-enable switch in the VCR won't let the deck erase over the tape. If you want to use the tape, place a piece of tape over the tab. Note that 8mm tapes have a slide tab. Push the tab over to prevent recording, and push it back to enable recording.

Sometimes the record-enable switch in the VCR becomes dirty or out of alignment. Check its operation with a meter. Visually

312

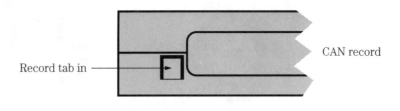

Record tab in

CAN record

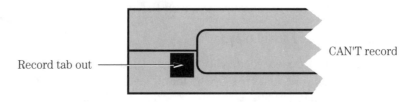

Record tab out

CAN'T record

Tape (as seen from back spine)

■ **9-26** *Only with the record tab in place can the VCR be placed in RECORD mode.*

examine the operation of the switch when you insert a tape into the deck. When the record tab is in place, the switch should be actuated (VHS and Beta). If the switch is not activated, it could be out of adjustment. Loosen the mounting screw(s) and readjust.

## Color sometimes flashes off and on

When you are watching a tape, does the color sometimes blank out for a moment, then come back on? Although this problem might be rooted in the anticopy signals encoded on the tape, it also can indicate a worn or flattened idler tire or pinchroller. Inspect the idler tire and pinchroller for excessive wear, grease build-up, glaze, or cracking. Replace or rejuvenate the rubber, as explained in chapter 5, "General cleaning and preventive maintenance."

A flattened pinchroller, like that in Fig. 9-27, will cause the color to come and go at a fairly quick rate, but the repetition will vary depending on the playback speed of the tape (the tape will almost flicker in SP speed but slowly pulse in EP speed). Visually check the roller for a flattened end and replace it.

A maladjusted television set is another cause of flaky color reception. Be sure you are fine-tuned to the output channel of the VCR. In rare cases, the VCR itself needs adjustment. One or more internal controls determine the color output intensity and timing. Refer to the service manual for your deck for more information.

Good pressure roller                Flattened pressure roller

■ **9-27** *A good pressure roller and a flattened pressure roller.*

## Weak video and audio playback

Magnetized heads can diminish the output strength of video and
audio heads. Most VCR manufacturers now recommend against
demagnetizing the video heads, but the audio heads should be
routinely demagnetized using a demagnetizing tool. For video
heads, almost all VCR makers suggest you use a demagnetizing
tool rated specifically for video work, not a regular audio deck
tool. Demagnetizers for audio machinery are often too powerful
for use inside VCRs.

Nevertheless, video heads do become magnetized, requiring de-
magnetization. Improper use of the demagnetizer tool can shatter
the video heads, so be sure you are careful and go slowly. Read
chapter 5, "General cleaning and preventive maintenance" for more
information on using demagnetizers.

## Sparkles in playback

Most VCRs made since the early 1980s use an electrostatic discharge element on the video heads (see Fig. 9-28). This element
helps neutralize the static that can build up in the head as the tape
rubs past it. If this discharge element is missing, broken, or not
making sufficient contact, the video playback can be riddled with
flashes of light. Recordings might not be as good as they can be,
and if your deck is a hi-fi model, the sound might crackle and pop.

Examine the discharge element and make sure it is securely
mounted. The tip of the element should touch (or barely touch)
the center of the video head drum. If there is considerable distance between the tip of the element and drum, use a pair of pliers
to bend the metal so better contact is made.

## Beta misthreading

Beta decks use a novel *threading ring* to grab the tape and press
it against the video and audio heads. A main roller brings the tape

■ **9-28** *Electrostatic discharge unit on top of the head drum.*

out of the cassette and wraps it around the head. When this roller is worn or becomes misadjusted, threading problems can occur. Check the condition of the main roller (also called the lead *guidepost*) and adjust or replace it as necessary.

### Bent or broken coil brackets

In Beta VCRs, the end-of-tape sensors are coils mounted on brackets on either end of the tape transport area. The coils work by sensing the presence of a magnetic flux, created by the coating on videotape. When the flux disappears, as it does when the aluminum strip at the ends of the tape are encountered, the VCR stops playing.

Because the coils must be in proximity to the tape to work, misalignment can sometimes cause the deck to prematurely turn off. Check the coil brackets to make sure that they are not bent or broken. If a bracket is broken, you can temporarily correct the problem by taping a small metal washer to the coil. Do not make this a permanent fix because you lose the safety device of automatic shutoff at the end of the tape.

## Tape counter skips or doesn't work

A counter is used to show the relative position of programming on the tape. On older Beta and VHS decks, the counter is mechanical, connected to a belt driven off the take-up reel. On all newer VCRs, the counter is electronic and usually borrows the sensor built into the take-up spindle. When the spindle turns, an infrared lamp and detector sense the rotation and increment or decrement the electronic counter.

In the mechanical system, the counter can stop working if the belt becomes worn or broken. Replace the belt and all should be fine. Almost all tape counters have a RESET button. The button is spring loaded and when pushed, resets the digits to zero. Sometimes, usually through abuse, the spring in the counter can get broken and the counter will no longer reset. You can try dismantling the counter and replacing the spring (a suitable replacement might be found at the hardware or hobby store) or you can replace the entire counter.

Failures in electronic counters also can be accompanied by other problems. For example, if the sensor system for the take-up reel is broken, the VCR will automatically shut off after 5 to 10 seconds when it realizes that the spindle isn't moving.

If the sensor is indeed working, look to shorts or opens in the wiring leading from the sensor to the counter. Use your ohmmeter to check for trouble. Also check the RESET button by the counter. Like mechanical counters, the button used with electronic counters also has springs. If the spring is broken, the button contacts can be jammed. That causes the counter to remain on "0000," even when the tape moves.

On a few VCRs, the electronic counter is driven by a separate sensor module. The module is what's technically called a shaft encoder and is like the sensor used in the take-up reel spindle. When the encoder turns, an infrared LED and sensor detect movement and direction, as shown in Fig. 9-29. This information is supplied to the counter electronics. The encoder is driven by the take-up reel, so check the belt or roller that connects them. Clean or replace as necessary.

Also check the sensor to make sure that it is outputting a signal. You can use a logic probe for this or a meter. Connect the encoder sensor as shown in Figs. 9-30 and 9-31, depending on whether you are using a logic probe or meter. Remember that most meters do not handle voltage changes very quickly, so manually turn the encoder shaft slowly with your fingers.

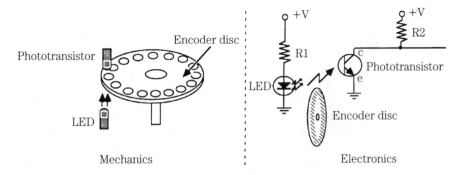

■ **9-29** *How a shaft encoder works. The disc might have holes or slits in it or can contain reflective strips (the LED and phototransistor are placed on the same side).*

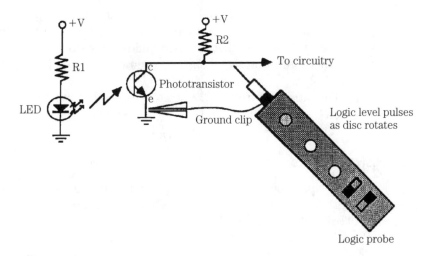

■ **9-30** *Connect the logic probe to the output (usually collector) of the phototransistor and watch for pulses as shaft encoder rotates.*

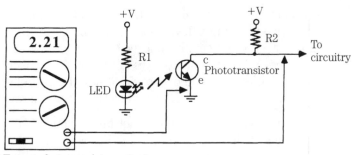

■ **9-31** *Connect the meter to the LED or phototransistor to test for voltage levels.*

## Picture flags

Have you ever noticed an unusual bending at the top of the screen? Video technicians usually refer to this phenomenon as *flagging*. It has many causes, some of which you can't correct:

☐ The tape tension is wrong. If the tape tension is too tight or too loose, the picture might flag at the top. Examine the supply reel back tension control and brake and try a new tape.

☐ The tape is recorded with an anticopying signal. The first anticopying techniques altered the vertical timing signals required by a VCR to lock onto the picture during recording. Unfortunately, even when you are not copying a tape, some TVs exhibit flagging due to timing instability problems. The loss of good vertical sync also upsets the horizontal synchronization that follows. This problem is most prevalent in American-made television sets manufactured before 1980. A video stabilizer that totally rebuilds the vertical synchronization signal has been known to help solve this problem.

☐ The tape has stretched. A stretched tape can exhibit flagging on the top of the screen. Also, the audio and overall picture can be bad.

**318**

## Excessive dropouts

In this instance, the term *dropout* has nothing to do with itinerant youth, but it involves a tape that is losing its magnetic coating. Loss of oxide results in a partial picture loss that appears on the screen as a white or black fleck. Dropout is mostly compensated for by the VCR's built-in dropout compensator, but a heavy amount of dropout, caused by relatively large areas of the tape with missing oxide coating, cannot be corrected.

Dropout is most prevalent at the beginning of tapes, where tapes are played the most. If the start of a tape is severely distorted by dropout, consider clipping out the offending portion and shortening the tape. Follow the procedures in chapter 5 on splicing videotape. Tapes that are very old or those that have been subjected to heavy use might not be salvageable and will have to be thrown away.

Avoiding excessive dropout is not only a good aesthetic practice, but it also helps prevent the video heads from becoming clogged. The more oxide coating that comes off the tape, the more material there is to dirty the video heads.

## Battery problems

Portable VCRs and camcorders operate on batteries (except when used with an adapter, they use household current). Camcorder and portable VCR batteries are usually rechargeable (the nickel-cadmium, or ni-cad, type) and are subject to certain difficulties.

☐ Ni-cad batteries should be fully charged before use, or their life expectancy is reduced. Before using the camcorder or portable, fully recharge the battery for the recommended time.

☐ Ni-cads are subject to a phenomenon known as *memory effect*. The battery tends to "remember" the duration of service previously expected of it, so if you regularly use only 50 percent of the battery capacity, battery life will be severely reduced. Whenever possible, "use up" the battery until it is completely dead. One way to do this is to leave the portable or camcorder on and let the battery run dead.

☐ Batteries should be regularly recharged, even if they are not used. Recharge batteries every two to four months, or they can be permanently damaged.

☐ Some camcorder battery packs use lead-acid or gelled electrolyte batteries. These, like ni-cads, are sealed against leakage, but given the right circumstances all batteries can leak. When not in use, remove the battery and store it in a cool, dry place.

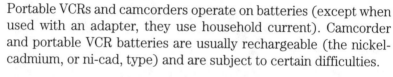

*319*

If the portable VCR or camcorder is not receiving battery power, check to make sure that the battery is properly charged. You can test the voltage of a battery to see if it is working properly, but this test might not be effective unless the battery is currently in use. Insert the battery in the VCR or camcorder and turn the unit on. Use your meter to measure the voltage output of the battery. It should be roughly the same as the rated voltage.

Check the battery terminals to make sure they are not corroded or broken. Clean the terminals if they are dirty or corroded; heavy corrosion or breakage must be repaired by replacing the terminals or battery compartment module. Use your meter to check the wiring from the battery compartment to the main board on the VCR or camcorder. You should get a reading all the way to the ON-OFF switch.

# Repairing camcorders

10

CAMCORDERS ARE COMBINATION VIDEO CAMERAS AND VCRs slimmed down to a size and weight you can easily carry. Camcorders contain a fully functional VCR that requires much of the same maintenance and repair as a full-sized deck. You must use caution, however, when working on camcorders because their small profile makes even routine maintenance and repair difficult.

## Disassembly

When at all possible, avoid disassembling your camcorder. Most camcorder models use "hidden" screws as a measure to discourage users from taking them apart (most still-camera manufacturers have used this scheme for years). The design of camcorders entails miniaturization and special construction techniques. Many parts are held in place by pressure or springs, and opening the case of the camcorder might release these tiny parts. However, if you must disassemble the camcorder, follow these steps:

321

☐ If possible, get a copy of the service manual. The manual will provide instructions on how to disassemble the unit and contains tips for keeping all the parts together. Finally, the manual will contain an assembly or exploded diagram that helps when putting the camcorder back together again.

☐ Lacking a service manual, disassemble the camcorder by carefully removing the exposed screws on the exterior of the case. If loosening a screw doesn't seem to loosen the case, temporarily retighten it and try another. Some screws on the exterior of the camcorder anchor internal parts, and these should not be removed.

☐ Screws are often hidden under the peel-away operating panels or the vinyl cushioning used on most camcorder models. Carefully peel away the panel or cushion to expose these hidden screws. Look also in or on the grip or battery compartment.

☐ Place the screws and all other parts you remove from the camcorder in a tray. Keep the parts separate to facilitate reassembly. It is crucial that you reassemble the camcorder using the same parts. Don't mix up the screws, washers, and other hardware. If necessary, take notes during disassembly.

☐ When all the screws are removed, carefully lay the camcorder on the table and separate the outer casing (or cover) from the main body. Go slowly; note the location of all parts that come loose.

☐ Some camcorders require you to loosen or remove the tape-loading cover before the case can be completely removed. The cover is generally held in place by two screws.

☐ Wires might attach the camcorder body to the case (particularly at the battery and operating control panels). Do not strain these wires by pulling the case too far from the camcorder. If wires tether the case to the camcorder, see if they can be removed at the connections. If not, carefully lay the case beside the camcorder body and work around it.

## Head cleaning

Because of the difficulty in disassembling and reassembling a camcorder, it's often better to clean the heads using a cassette cleaner. Just about any wet or wet/dry cleaner will do; experiment with a few brands until you find a cleaning system that works best.

Use the head-cleaning tape as recommended by the manufacturer. After using the tape, wait several minutes for the cleaning fluid that has been deposited inside the camcorder to evaporate completely. Inserting and playing a tape when the unit is still damp can reclog the heads and can cause the tape to stick.

A more thorough head cleaning entails removing the cover of the camcorder and using swabs dipped in alcohol or other cleaning agent. Follow the head-cleaning procedures in chapter 5. Note that compact VHS-C camcorders have four or more video heads; be sure to clean them all. Many 8mm and some VHS camcorders also have flying erase heads built onto the head drum. Be sure to clean these as well.

All camcorders are the top-loading variety, which means that when the tape-loading cover is open (the unit is ready to accept a tape), the video-head drum is accessible. If possible, keep the case of the camcorder on and clean the video heads by using a sponge- or chamois-tipped swab, poking it through the opening in the tape-loading cover. Be sure that you don't apply excessive pressure on the heads. You can rotate the drum to access all of the heads by spinning the top part with your fingers or the eraser end of a pencil, but take care not to touch the heads with any instrument except the swab soaked in cleaning fluid.

How often should you clean the heads depends on how much you use the camcorder. You should clean the heads whenever your camcorder shows definite signs of dirty heads. You also should clean the heads after using the camcorder outdoors in adverse conditions—in dust, water spray, sand, and so forth. If you use your camcorder regularly, clean the heads about once every three to six months. For occasional use, clean once every 6 to 12 months.

## Dusting and general cleaning

Because camcorders are primarily used out of doors, they are especially likely to become soiled by dust, dirt, and other contaminants. At regular intervals, clean the exterior of the camcorder with a mild household cleaner, but remember to apply the cleaner to the sponge or towel—never spray the cleaner directly onto the camcorder because the excess might ooze inside.

Clean around the control switches with a cotton swab dipped in cleaner or alcohol. Avoid the use of petroleum-based solvent cleaners, especially acetone. These can remove paint and melt the plastic casing of the camcorder. If you have disassembled the camcorder, use a soft brush lightly to dust the interior of the unit. A can of compressed air can be used to rid dust and dirt from hard-to-reach places. If the sediment is heavy, use a cleaner/degreaser as described in chapter 5.

## Parts cleaning and replacement

Camcorders use slightly fewer parts than their full-sized cousins, but that doesn't mean that mechanical failure is less likely to occur. Rubber belts, rollers, and other components can fail after time and might need cleaning or replacement.

Generally, you clean and replace parts such as idlers, belts, and pressure rollers the same as you do regular VCRs. Because of their compact nature, however, a variety of novel schemes are used to attach components to the camcorder.

Tires and rollers might be secured with just a press fit, but the fit might be extremely tight, requiring a special tool for removal and replacement. On other camcorders, rollers and tires are secured using miniature jeweler's screws or hex screws. Be sure to use the proper-size tool or you risk stripping the head of the screws. If some screws seem overly tight, apply a very small dab of oil to the head and shaft area and wait for the oil to penetrate.

It's a joy to work on a camcorder that is designed so drive belts are out in the open and are readily accessible. However, the midget design of camcorders often precludes this, and belts are typically hidden under other components and are difficult to reach.

At least some portion of the belt is usually accessible, and you can use a pair of tweezers, surgical forceps, or dentist's picks to examine and—only if necessary—extract the belt for replacement. Note that removing the belt is the easy part; putting a new one on is only slightly less frustrating than getting the ball in the milk bottle at the county fair. Note that some components, such as motors and circuit boards, need to be removed in order to replace a belt.

## Camera care

The camera portion of the camcorder needs little maintenance beyond the following:

☐ Clean the lens prior to each use. Never clean the lens with a dry lens tissue; wet the tissue or lens with a suitable cleaner. If you have attached a clear UV (ultraviolet) or skylight filter over the front of the lens (to protect the actual camera lens from accidental injury), periodically remove the filter and clean it.

☐ Use a cotton swab dipped in alcohol to clean the lens barrel and camera controls, and blot dry.

☐ If the camera gets out of whack—the focus is thrown off or the image appears lopsided, for example—adjustment of some internal controls might be necessary. Do not attempt to adjust any internal controls without the assistance of the service manual. The manual provides test points and procedures for aligning the camera section for proper picture taking. Note that the latest camcorders use solid-state imaging devices, which rarely require the adjustments needed by vacuum-tube pickups.

# How to help your camcorder live longer

Camcorders aren't inexpensive, even the cheap ones aren't cheap. Yet among the prized possessions in your video den, your camcorder might be treated with the least respect.

You unceremoniously throw the camcorder in the trunk of the car, you pull it out at the motorcross race, never giving the plumes of dust in the air a second thought; you put in tape after tape after tape, no matter how much grime is building up inside the poor thing.

All these habits—and others—take their toll on the life of your camcorder. A camcorder that might live a healthy, productive life of six or seven years might be a broken-down derelict after only a year or two. Even a bargain-basement $500 camcorder deserves more than that.

Here's how to extend the life of your camcorder beyond its one year warranty period, without having to become a fastidious Felix Unger. You'll learn basic things you can do to keep your camcorder healthy. Plus you'll find suggestions for simple cleaning and routine maintenance that can literally double or triple the life of your camcorder, while keeping it out of the camcorder-doctor's office.

# Camcorder care and feeding

Since grade school you are told to eat right, exercise regularly, and refrain from bad habits, like smoking, drinking, and watching reruns of "Gilligan's Island." A healthy body to begin with helps us live longer and avoid illness.

It's the same with camcorders. Treat your camcorder right from the day you buy it, and odds are it will live longer and won't break down as much. Even unintentional abuse of a camcorder can slowly erode its immune system, requiring either replacement or expensive repairs. Here is a short laundry list of simple steps you can take to keep your camcorder fit as a fiddle.

☐ When not using your camcorder, stash it away so it won't get dusty. If you don't have a special place for it, how about in a corner of your sock drawer. Wrap the camcorder in a towel to keep it from being knocked around in the drawer.

☐ Keep your camcorder clean at all times. If you see dust or dirt on it, wipe it off, with a handkerchief (preferably a clean one). Don't let your camcorder become so dirty that you need to get out the liquid cleaner. If you ever do have to use a liquid cleaner, apply a small amount of it to the cleaning cloth; never spray the cleaner directly on the camcorder.

☐ When not using your camcorder, always disconnect the battery and store it separately. Removing the battery prevents costly damage to the camcorder if the battery should ever leak. See the section "Proper care and use of camcorder batteries" for more tips on taking care of batteries.

☐ Invest $10 in a UV filter for the lens of your camcorder. The filter, which is essentially a piece of glass with a coating on it that absorbs most ultraviolet light, screws onto the front of the lens. It's cheaper to replace the UV filter should it get scratched rather than the entire camcorder lens. UV filters are available in different sizes at photo stores.

☐ Clean the lens with a lens brush, available at any photo store. The kind with the built-in blower is okay. NEVER use lens cleaning tissue designed for eyeglasses to clean the camcorder lens. Most eyeglass cleaners have chemicals in them that can eat away the coating on your lens!

## Don't forget the supplies and accessories

Don't forget that your camcorder supplies and accessories need proper care, too.

Store videotapes in a cool, dry place. Keep tapes away from dust and grime, or else you'll bring the junk inside the camcorder whenever you insert a dirty tape. If you have an external microphone, protect the sound element against water, dirt, and shock. When not in use store it in a case, or wrap it up in a sock. If the microphone has its own battery, check it every so often. Remove the battery if you won't be using the microphone for a while.

Most video lights use tungsten-halogen light bulbs, which are sensitive to shock, especially when they are warm. Keep the lights in a case when not in use. If the bulb is exposed in the light fixture (there's no glass cover over it), stuff a rag over the bulb to protect it (wait until the bulb is cool first!). Wipe fingerprints from the bulb; the oil from your skin can reduce the life of the bulb.

## Proper care and use of camcorder batteries

Among the common ailments of camcorders is a bad battery. Camcorder batteries use rechargeable nickel-cadmium (ni-cad) cells, which are normally durable and trustworthy. Still, they aren't infallible, and even the best camcorder battery can go completely dead. They've even been known to die and take the camcorder with them. So be wary.

Always, always store camcorder batteries in a cool, dry place. Avoid putting a battery in a hot car, even if it's only for an hour. Not only does this rob some of the charge from the battery, it can weaken the battery and reduce battery life. If you don't plan on using your camcorder for a while, discharge the battery by turning on the camcorder and keeping it on (you might need to play through a tape to keep the camcorder from automatically shutting off, if it has a power-saving feature). Ni-cad batteries last longer if they are stored in a moderately discharged state. However, never let a ni-cad battery discharge too much. Otherwise, the cells inside the battery pack might be damaged.

Like all batteries, ni-cads can leak. So, when not using your camcorder, remove the battery pack and store it separately. Signs of a leaked battery: a white powder that oozes from the seams of the pack. This powder is highly caustic, and can burn your skin. Dispose of the battery pack properly (it's ruined), wash your hands thoroughly, and clean anything that came into contact with the powder.

A teeny bit of corrosion of the battery terminals on a camcorder battery pack is normal, especially if you forget to remove the battery pack from the camcorder for storage. You can remove the corrosion by cleaning the electrical contacts of both the battery and camcorder with a pencil eraser. Be sure not to leave any eraser bits behind.

Modern ni-cads don't exhibit as much memory effect that plagued earlier versions. Still, you will want to get into the habit of using your battery until it is discharged (the LOW-BATT light comes on inside your camcorder). Recharge the battery then. Letting the battery discharge before recharging will allow it to last longer between charges.

## Special tips when traveling with a camcorder

Nowadays when people get ready for their vacation, they pack their camcorder before their shaver and hair dryer. That's how important this little trinket has become. Most of the serious damage that can happen to a camcorder is when traveling, so keep these important points in mind the next time you pick your camcorder as your traveling mate.

**Avoid packing your camcorder as luggage. Instead, carry it on board the plane.** Sure, it will take you a bit longer to get through the airport security checkpoint, but your camcorder investment is worth it. Camcorders are probably stolen more out of luggage than anything else, and besides, the luggage handlers at most airports aren't known for their gentle hands.

**Always pack your camcorder in something; never carry it on board nude.** If you have a separate case for your camcorder, so much the better. The case will protect the camcorder from the normal bumps and grinds encountered during travel. If you don't have a camcorder case, wrap it up in a beach towel and tuck it in the middle (not along the sides!) of your suitcase.

**The sun, sand, and water are great vacation spots for people, but awful for camcorders.** If you take your camcorder with you on an outdoor outing, be sure to keep it away from harsh elements. If you plan on using the camcorder around the water, consider buying a water resistant cover for it (it doesn't have to be one of the expensive underwater housings). In a pinch you can use a heavy-duty refrigerator storage bag. Get one big enough to hold the camcorder. You can point the lens out of the opening of the bag, and operate the camcorder through the plastic.

Finally, if you're coming in from the rain or cold, wait a few minutes for your camcorder to warm up. Going from cold to warm can cause condensation inside the camcorder, which can cause problems.

# Sources

## VCR manufacturers

Aiwa America, Inc.
800 Corporate Dr.
Mahwah, NJ 07430
(201) 440-5220

Canon USA
One Canon Plaza
Lake Success, NY 11042
(516) 488-6700

Chinon America
43 Fadem Rd.
Springfield, NJ 07081
(201) 376-9260

Daewoo International
America Corp.
100 Daewoo Place
Carlstadt, NJ 07072
(201) 935-8700

Emerson Radio Corp.
One Emerson Ln.
North Bergen, NJ 07047
(201) 854-6600

Fisher Corp. (Sanyo Fisher
USA Corp.)
21350 Lassen St.
Chatsworth, CA 91311
(818) 998-7322

GE
(See Thomson Consumer Electronics)

Goldstar Electronics, Inc.
1000 Sylvan Ave.
Englewood Cliffs, NJ 07632
(201) 816-2000

Go-Video Inc.
14455 N. Hayden Rd., Ste 219
Scottsdale, AZ 85260
(602) 998-3400

Hitachi Home Electronics
401 W. Artesia Blvd.
Compton, CA 90220
(213) 537-8383

Instant Replay
2951 S. Bayshore Dr.
Coconut Grove, FL 33133
(305) 448-7088

JVC Company of America
41 Slater Dr.
Elmwood Park, NJ 07407
(201) 794-3900

Kenwood USA Corp.
2201 E. Dominguez St.
P.O. Box 22745
Long Beach, CA 90810-5745
(213) 639-9000

Matsushita Electric Corp. of America
(See Panasonic)

Mitsubishi Electronics of America
800 Biermann Ct.
Mount Prospect, IL 60056
(708) 298-9223

Montgomery Ward
535 W. Chicago Ave.
Chicago, IL 60610

Panasonic Co.
One Panasonic Way
Secaucus, NJ 07094
(201) 348-7000

Radio Shack
1300 One Tandy Center
Fort Worth, TX 76102
(817) 878-4852

RCA
(See Thomson Consumer
Electronics)

Ricoh Corp.
180 Passaic Ave.
Fairfield, NJ 07006
(201) 882-2000

Samsung Electronics America, Inc.
301 Mayhill St.
Saddle Brook, NJ 07662
(201) 587-9600

Sansui USA, Inc.
1290 Wall Street West
Lyndhurst, NJ 07071
(201) 460-9064

Sanyo Fisher USA Corp.
21350 Lassen St.
Chatsworth, CA 91311
(818) 998-7322

Sears Roebuck & Co.
4640 Roosevelt Blvd.
Philadelphia, PA 19132

Sharp Electronics Corp.
Sharp Plaza
Mahwah, NJ 07430
(201) 529-8200

Shintom West Corp.
20435 South Western Ave.
Torrance, CA 90501
(213) 328-7200

Sony Corp. of America
1 Sony Dr.
Park Ridge, NJ 07656
(201) 930-1000

Tatung
2850 El Presidio St.
Long Beach, CA 90810
(213) 637-2105

Technics
One Panasonic Way
Secaucus, NJ 07094
(201) 348-7000

Thomson Consumer Electronics
(RCA and GE)
600 N. Sherman Dr.
Indianapolis, IN 46201
(317) 267-5000

Toshiba America, Inc.
82 Totowa Rd.
Wayne, NJ 07470
(201) 628-8000

Vector Research
1230 Calle Suerte
Camarillo, CA 93010
(805) 987-1312

Yamaha
6722 Orangethorpe Ave.
Buena Park, CA 90620
(714) 522-9105

Yashica
100 Randolph Rd.
Somerset, NJ 08875
(201) 560-0060

Zenith Electronics Corp.
1000 Milwaukee Ave.
Glenview, IL 60025
(312) 391-8181

## Accessories manufacturers

Acme-Lite Manufacturing Company
3659 W. Lunt Ave.
Lincolnwood, IL 60645-1210
(312) 588-2776
*Camcorder batteries, cassette
rewinders, tripods, other video
accessories*

Allsop Inc.
4201 Meridian St.
P.O. Box 23
Bellingham, WA 98227
(206) 734-9090
*Head-cleaning supplies*

Ambico
50 Maple St.
Norwood, NJ 07648
(201) 767-4100
*Cleaning supplies, cables, etc.*

Arkon
11627 Clark St., Suite 101
Arcadia, CA 91006
(818) 357-1133
*Video lights, camcorder and VCR
accessories*

AudioControl
22313 70th Ave. W
Mountlake Terrace, WA 98043
(206) 775-8461
*Cables, speakers, and audio
processors*

BASF
Crosby Dr.
Bedford, MA 01730
(617) 271-4000
*Videotape*

Battery Tech, Inc.
28-25 215 Place
Bayside, NY 11360
(718) 631-4275
(800) 442-4275
*Replacement batteries*

Bib America
10497 Centennial Rd.
Littleton, CO 80127
(303) 972-0410
*Cleaning supplies, cables,
accessories, etc.*

BP Electronics
260 Motor Parkway
Hauppauge, NY 11788
(516) 435-8777
*Cleaning supplies, accessories*

Bush Industries
One Mason Dr.
Jamestown, NY 14702
(716) 665-2000
*Cabinets*

Calibron, Inc.
2950 Lake Emma Rd.
Lake Mary, FL 32746
(407) 323-2400
*VCR accessories, cleaners,
videotape*

**331**

Claggk Inc.
P.O. Box 4099
Farmingdale, NY 11735
*VCR cleaning accessories and
ISCET VCR Cross Reference
guide*

Discwasher
4310 Transworld Rd.
Schiller Park, IL 60176
(312) 678-9600
*Head-cleaning supplies*

Everquest
875 S. 72nd
Omaha, NE 68114
(402) 554-0383
*Channel converters, video signal
broadcasters, video enhancers*

Fuji Photo Film USA
555 Taxter Rd.
Elmsford, NY 10523
(914) 789-8100
*Videotape*

GC Electronics (GC-Thorsen)
P.O. Box 1209
Rockford, IL 61105-1209
(815) 968-9661
*TV and VCR accessories, cleaners*

Gemini Industries, Inc.
215 Entin Rd.
P.O. Box 1115
Clifton, NJ 07014
(201) 471-9050
*Video accessories, antennas,
videotape*

Gemstar Development Corp.
135 N. Los Robles, Suite 870
Pasadena, CA 91101
(818) 792-5700
*VCR Plus+ VCR programmer*

Gusdorf Corp.
11410 Lackland Rd.
St. Louis, MO 63146
(314) 567-5249
*Cabinets*

HTS, Inc.
90 Inverness Circle East
P.O. Box 6552
Englewood, CO 80155
(303) 790-4445
*Video signal broadcasters,
programmable remote control*

Infrared Research Labs, Inc.
820 Davis St.
Evanston, IL 60201
(708) 328-8043
*Universal remote controls*

Jasco Products Co., Inc.
311 N.W. 122nd
P.O. Box 466
Oklahoma, OK 73114
(405) 752-0710
*Video accessories, cables, video
rewinders*

Kinyo Co., Inc.
14235 Lomitas Ave.
La Puente, CA 91746
(818) 333-3711
*Video rewinders, tripods*

Maxell Corp. of America
22-08 Rte. 208 S.
Fair Lawn, NJ 07410
(201) 794-5900
*Videotape*

Memtek/Memorex
P.O. Box 901021
Fort Worth, TX 76101
(817) 878-6700
*Videotape*

Mini-Vac, Inc.
217 S. Orange St., #4
Glendale, CA 91204
(818) 244-6777
*Hand-held portable vacuum
cleaner*

Monster Cable Products Inc.
274 Watis Way
South San Francisco, CA 94080
(415) 871-6000
*Cable*

Multi-Video Inc.
P.O. Box 35444
Charlotte, NC 28235
(704) 536-6928
*Videotape repair supplies and
instruction*

O'Sullivan Industries
1900 Gulf St.
Lamar, MO 64759
(417) 682-3322
*Cabinets*

Polaroid Corp.
549 Technology Square
Cambridge, MA 02139
(617) 577-2000
*Videotape*

Philips Consumer Electronics Co.
P.O. Box 555
Jefferson City, TN 37760
(615) 457-3801
*VCR/tape care products, video
accessories*

Rabbit Systems
100 Wilshire Blvd.
Suite 600
Santa Monica, CA 90401
(213) 393-9830
*In-home RF video links*

Recoton Corp.
46-23 Crane St.
Long Island City, NY 11101
(718) 392-6442
*Accessories*

TDK Electronics Corp.
12 Harbor Park Dr.
Port Washington, NY 11050
(516) 625-0100
*Videotape*

3M Consumer Video and Audio
Products
3M Center
Building 223-5N-01
St. Paul, MN 55144
(612) 736-3544
*Videotape, VCR cleaning
accessories*

Showline Video Corp.
120 Beacon St.
Boston, MA 02116
(617) 262-6844
*Video processors*

Sima Products Corp.
8707 Skokie Blvd.
Skokie, IL 60077
(708) 679-7462
*VCR accessories, light/audio
equipment and accessories, VCR
cleaning accessories*

Solidex, Inc.
615 W. Allen Ave.
San Dimas, CA 91773
(714) 599-2666
*Videotape rewinders, accessories*

Vidicradt, Inc.
8770 S.W. Nimbus Ave.
Beaverton, OR 97005
(503) 626-1918
*Video processors*

Vision Perfect Corp.
4265 San Felipe, #820
Houston, TX 77027
(713) 621-1695
*VCR cleaning accessories,*
*videotape rewinders*

## Sources for VCR parts and test equipment

Alfa Electronics
741 Alexander Rd.
Princeton, NJ 08540
(609) 520-2002
(800) 526-2532
*Test equipment*

All Electronics Corp.
P.O. Box 567
Van Nuys, CA 91408-0567
(818) 904-0524
(800) 826-5432
*Electronic parts and test*
*equipment*

Andrews Electronics
25158 Stanford Ave.
Valencia, CA 91355
(800) 274-4666
*VCR replacement parts*

C & S Sales, Inc.
1245 Rosewood
Deerfield, IL 60015
(708) 541-0710
(800) 292-7711
*Test equipment*

Circuit Specialists, Inc.
P.O. Box 3047
Scottsdale, AZ 85271-3047
(800) 528-1417
(602) 966-0764
*Test equipment, replacement*
*electronic parts, cleaning*
*supplies*

GMB Electronics Supply
140 Terminal Rd.
Setauket, NY 11733
(800) 874-1765
(516) 689-3400
*VCR replacement parts*

Halted Electronic Supply
2500 Ryder St.
Santa Clara, CA 95051
(408) 732-1573
*Electronic parts and test*
*equipment*

MCM Electronics
650 Congress Park Dr.
Centerville, OH 45459-4072
(513) 434-0031
(800) 453-4330
*VCR parts (belts, replacement*
*heads, idler rollers), cleaning*
*supplies, and test equipment*

Mouser Electronics
2401 Hwy 287 North
Mansfield, TX 76063
(800) 992-9943
*VCR parts, electronic*
*replacement parts, test*
*equipment*

Optoelectronics, Inc.
5821 N.E. 14 Ave.
Ft. Lauderdale, FL 33334
(305) 771-2050
*Test equipment*

Parts Express
340 E. First St.
Dayton, OH 45402
(513) 222-0173
*Repair parts, infrared detector*
*pen, cleaning supplies*

**334**

R & S Surplus
1050 E. Cypress St.
Covina, CA 91724
(818) 967-0846
*Surplus test equipment*

Tandy Consumer Parts
7439 Airport Freeway
Fort Worth, TX 76118
(800) 243-1311
*Replacement parts for Radio
Shack/Realistic VCRs, plus OEM
brands*

Tritronics
1306 Continental Dr.
Abington, MD 21009
(800) 638-3328
(301) 676-7300
*VCR replacement parts*

Tucker Electronics and Computers
16717 Reserve St.
Garland, TX 75042
(214) 348-8800
(800) 572-4642
*Test equipment*

Union Electronic Distributors
16012 Cottage Grove
South Holland, IL 60473
(800) 648-6657
(312) 333-4100 or (312) 468-7300
*VCR replacement parts*

Western Test Systems
530 Compson St., Unit C
Broomfield, CO 80020
(303) 438-9662
(800) 538-1493
*Test equipment*

Windward Products
P.O. Box 387
Moffett Field, CA 94035
(800) 470-9463
*Test equipment*

## Miscellaneous

Electronic Industries Association
2001 Pennsylvania Ave. NW
Washington, DC 20006-1813
(202) 457-4919
*Trade association for the
electronics industry*

Heathkit Educational Systems
(800) 44-HEATH
*Self-study courses on VCR
servicing*

International Society of Certified
Electronics Technicians
2708 W. Berry St.
Fort Worth, TX 76109
(817) 921-9101
*Certification group for
electronics technicians;
publisher of VCR Cross Reference*

School of VCR Repair
6065 Roswell Rd.
Atlanta, GA 30328
(800) 223-4542
*Correspondence course for VCR
repair*

Society of Audio/Video Consultants
(SAC)
P.O. Box 10957
Beverly Hills, CA 90213
(213) 550-8889
*Trade association for A/V
consultants, publications*

# Further reading

INTERESTED IN LEARNING MORE ABOUT VIDEO AND VCRs? Here is a selected list of magazines and books that can enrich your understanding and enjoyment of the world of video.

## Magazines

*The Perfect Vision*
2 Glen Ave.
Sea Cliff, NY 11579
(516) 671-6342
For those who take video seriously. Includes reviews and articles on video hardware and software.

*Audio/Video Buyer's Guide*
Harris Publications
1115 Broadway
New York, NY 10010
Seasonal buyers' guide on home and car audio and video, available at newsstands. Light reading, but might be somewhat out-of-date by the time you get it. No subjective reviews of VCR hardware.

*Camcorder Report*
2660 East Main St.
Ventura, CA 93003
General-interest magazine on camcorder care and use.

*Nuts & Volts Magazine*
P.O. Box 1111
Placentia, CA 92670
Classified-ad magazine aimed at the electronics enthusiast. Ads (classified and display) of interest to VCR buffs. Good source for new and used parts and test equipment.

*Popular Electronics*
Electronics Now
500 B-County Blvd.
Farmingdale, NY 11735
(516) 293-3000
Monthly magazines with projects and articles for the electronics hobbyist. Occasional technical article on VCRs and video.

*Video Magazine*
460 W. 34th St.
New York, NY 10001
Monthly magazine on all types of video items. Includes hardware and software reviews, technical articles (how it works), buyers' guides, and how-to stories. VIDEO also regularly publishes special buyer's guide issues.

*Videomaker Magazine*
1370 Dell Ave.
Capbell, CA 95008
(408) 866-8300
Monthly magazine devoted mostly to the technical and artistic use of camcorders. Also includes occasional camcorder maintenance articles.

## Books

*How to Read Electronic Circuit Diagrams—2nd Edition*
Robert M. Brown, Paul Lawrence, and James A. Whitson
TAB Books, Catalog #2880
How to read and interpret schematic diagrams.

*Maintaining and Repairing VCRs—2nd Edition*
Robert L. Goodman
TAB Books, Catalog #3103
Technical discussion on the repair and maintenance of VCRs. Aimed at the service technician.

*Troubleshooting and Repairing Camcorders*
Homer L. Davidson
TAB Books, Catalog #3337
In-depth discussion of repair and maintenance of camcorders. Technical in nature, with representative schematic diagrams.

**338**

*VCR Cross Reference*
International Society of Certified Electronics Technicians
(ISCET)
2708 W. Berry St.
Fort Worth, TX 76109
(817) 921-9101
Extensive cross-reference listing of VCR components.

# Soldering tips and techniques

SUCCESSFUL REPAIR OF YOUR VCR DEPENDS ON HOW WELL you can solder two wires together. Soldering sounds and looks simple enough, but there really is a science about it. If you are unfamiliar with soldering or want a quick refresher course, read this short soldering primer.

## Tools and equipment

Good soldering requires the proper tools. If you don't have them already, they can be purchased at Radio Shack or almost any electronics store.

### Soldering iron

You'll need a soldering iron, of course, but not just any old soldering iron. Get a soldering "pencil" with a low-wattage heating element. For electronics work, the heating element should not be higher than about 30 W. Most soldering pencils are designed so you can change heating elements as easy as changing a light bulb.

Do not use the instant-on type soldering guns, favored in the old tube days. They create far too much unregulated heat, and they are too large to solder most joints on a PCB effectively.

If your soldering iron has a temperature control and readout, dial it to between 665° and 680°. This setting provides maximum heat with the minimum danger of damage to the electronic components. When you are not using your soldering iron, keep it in an insulated stand. Don't rest the iron in an ashtray or precariously on the carpet. You or some precious belonging is sure to be burned.

### Soldering tip

The choice of soldering tip is important. For best results, use a fine tip designed specifically for printed circuit board use. Tips are made to fit certain types and brands of heating elements, so make sure you get the type for your iron. If the tip doesn't come pretinned, tin it by attaching the tip to the iron, allow it to heat up, and after the iron is hot, apply a thin coat of solder to the entire tip.

### Sponge

Keep a damp sponge by the soldering station and use it to wipe off extra solder. You'll have to rewet the sponge now and then if you are doing lots of soldering.

### Solder

Use only rosin-core solder. It comes in different thicknesses; for best results, use the thin type (0.050 inch) for PCB work. Never use acid-core or silver solder on electronic equipment.

### Soldering tools

Basic soldering tools include a good pair of small needle-nosed pliers, tweezers, wire strippers and wire cutters (sometimes called side or diagonal cutters). The stripper should have a dial that lets you select the gauge of wire you are using. A pair of "nippy" cutters that cut wire leads flush to the surface of the board are handy but not absolutely essential.

### Cleaning supplies

After soldering and when the components and board are cool, you should spray or brush on some flux remover. Isopropyl alcohol also can be used for cleaning.

### Solder vacuum

A solder vacuum is a suction device that is used to pick up excess solder. It is often used when desoldering (removing a wire or component from the board). Solder can also be removed using a length of copper braid. Most electronics stores sell spools of it specifically for solder removal.

## Basic soldering

The basis of successful soldering is that the soldering iron is used to heat the work, whether it is a component lead, a wire, or whatever. You then apply the solder to the work. Do not apply solder directly to the soldering iron. If you take the shortcut by melting the solder on the iron, you might end up with a cold solder joint. A cold joint doesn't adhere well to the metal surfaces of the part or board, so electrical connection is impaired.

Once the solder flows around the joint (and some will flow to the tip), remove the iron and let the joint cool. Avoid disturbing the solder as it cools; a cold joint might result. Do not apply heat any longer than necessary. Prolonged heat can permanently ruin electronic components. A good rule of thumb is that if the iron is on any one spot for more than 5 seconds, it's too long.

If possible, keep the iron at a 30° to 40° angle to the work, as shown in Fig. C-1. Most tips have a beveled tip for this purpose. Apply only as much solder to the joint as is required to coat the lead and circuit-board pad. A heavy-handed soldering job might lead to *soldering bridges*, a soldered joint that melds with other joints around it. At best, solder bridges prevent the circuit from working; at worst, they cause short circuits that can burn out the entire board.

## Replacing components

Removing a soldered component first requires that you remove all the solder holding it in place. Use the soldering iron to melt the joint, and as the solder flows, suck it up with a solder vacuum or wick. Re-

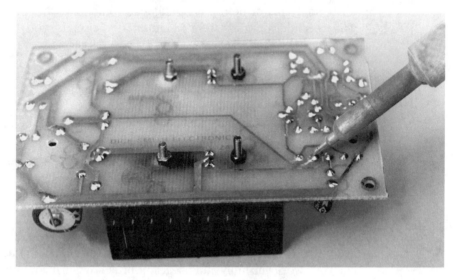

■ **C-1** *The proper technique for applying the soldering iron to a printed circuit board.*

move enough solder so the component lead is free. If you can't get all the solder up the first time, let the joint cool and try again.

Clean the old joint and the leads of the replacement component in alcohol. The alcohol removes any oil that can impair the grip of the joint after soldering.

Insert the new component gently. Don't pull on the lead or you can damage the component. Once the part is seated on the board, bend the leads slightly to keep it in place. Solder as usual. If you are resoldering a wire onto a terminal, wrap the stripped end of the wire around the eyelet of the terminal prior to soldering.

## A good solder joint

A good solder joint should be bright and shiny. A joint that looks dull is probably cold and should be resoldered. The joint should not have any sharp peaks. If you see peaks, the solder didn't flow well enough to make a good connection. Excess solder that forms on the tip (another cause for the peaks) should be removed using the damp sponge.

## Electrostatic discharge

Electrostatic discharge—better known as a *carpet shock*—can ruin electronic components. You should remove the excess static buildup from your body by touching some grounded metal object prior to soldering or before handling electronic parts and boards. If you are soldering transistors and integrated circuits, you should use a grounded soldering iron as well as an antistatic wrist band and table mat.

## Iron tip maintenance and cleanup

After soldering, let the iron cool. Loosen the tip from the heating element and store it for next use. After several soldering sessions, the tip should be cleaned using a soft brush. Don't file it or sand it down with emery paper.

Invariably, little nuggets of solder will be left around after a repair job. Make sure that these balls of once-molten solder balls are not left on the PCB or in the VCR cabinet. The solder might bridge wires or board traces together, causing a serious short circuit. Inspect your work carefully and use a soft brush to whisk away stray bits of solder.

# Attaching an F connector

YOU EITHER CAN BUY VIDEO CABLES CUT TO LENGTH WITH THE proper F connector already attached to them, or you can make the cables yourself. Making your own cables saves you money, and the cables can be cut to the proper length for the job. Figure D-1 on the next page is a step-by-step guide to attaching an F connector onto coaxial cable. Please note that F connectors are designed for a specific cable type; for example, connectors designed for RG59 cable won't fit on RG6 cable.

## What to do

1. Strip the outer jacket insulation from the cable to expose about $7/16$ inch of the braided wire. At this point, don't cut the braid.
2. Fan out the braid, fold it back over the cable jacket, and trim to about $1/8$ inch in length.
3. Some cables have an aluminum shield under the braid. Strip off the excess shield and insulation foam to expose $1/4$ inch of the center conductor—do not cut too deeply into the dielectric or you'll nick the center conductor. Make sure the center conductor is clean and appears bright.
4. Slide the F connector over the foam but under the braid. Push the fitting into the cable until it seats firmly and no braid is showing. If the fitting is the crimp type, crimp it with the proper tool.

## Tools list

Although specialty tools aren't absolutely necessary for attaching F connectors to coaxial cable, using the right tool makes the work go faster. The results are more professional, too.

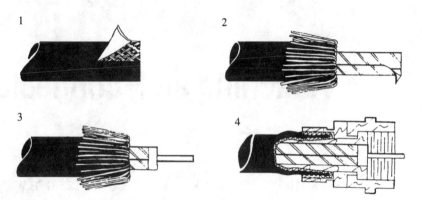

■ **D-1** *The steps in attaching an F connector to a coaxial cable.*

You can cut the foam insulation from around the center conductor with a knife or use a conventional wire stripper. The stripper helps prevent nicking the conductor. Perhaps the best method is to use a coax cable stripper tool. This stripper, which works with all types of television coax, automatically cuts off the proper amount of outer insulation and foam.

Because F connectors are large, they require a special tool for proper crimping (regular wire crimpers don't work well). The tool is available in a variety of styles—your choice from economy (as shown in Fig. D-2) to heavy-duty, professional models. Prices range from $2 to $12.

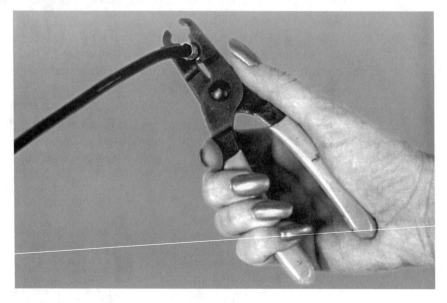

■ **D-2** *Use a crimper to securely attach the F connector to the cable.*

If your cables will be used exclusively indoors, you can use crimpless F connectors. These screw securely onto the end of the cable. Crimpless connectors are about twice as expensive as their crimped cousins, so you might not want to use them extensively.

# Television frequency spectrum

ALL TELEVISION BROADCAST CHANNELS FALL BETWEEN 54 MHz to 890 MHz. This comprises all VHF lowband, VHF highband, UHF (standard), UHF (translator), RCI subchannels, and midband and super-band cable television channels.

Table E-1 on the next page shows a frequency spectrum chart that details the allocation of television channels with adjacent service frequencies, including shortwave, FM radio, and ham radio. Table E-2, shown on pages 351-353 lists over-the-air TV channel assignments with audio, video, and color-carrier frequencies.

Note that the exact frequencies of the picture, color, and sound carriers might be different in cable systems using IRC and HRC frequency-offset broadcasting standards.

■ Table E-1 Television frequency spectrum chart.

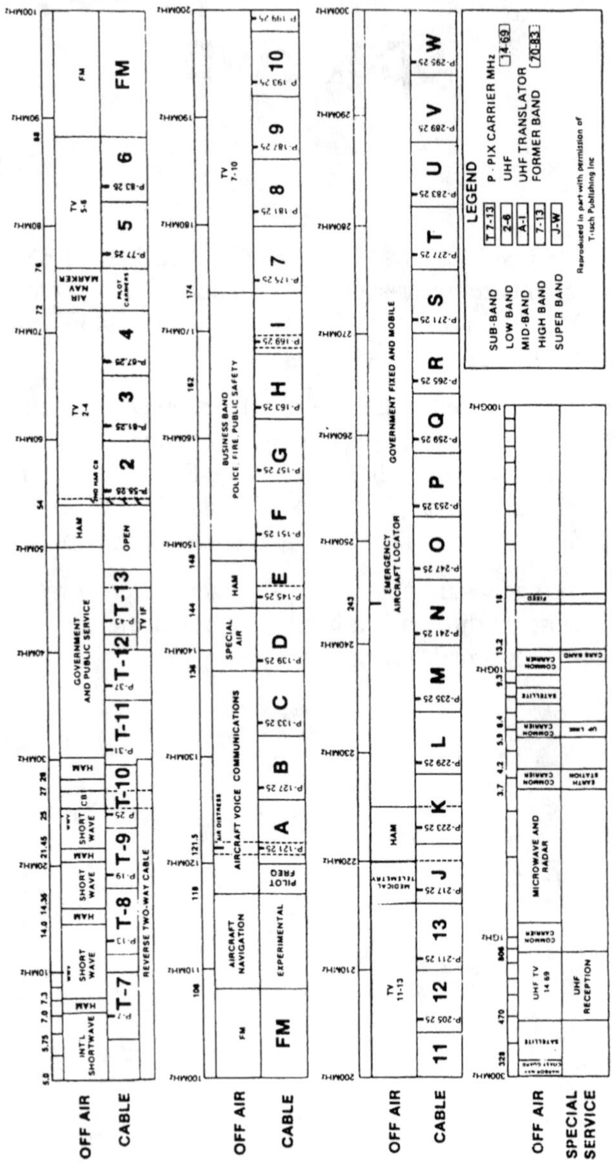

■ Table E-2 TV channel assignments.

| Channel no. | Frequency range (MHz) | Picture carrier (MHz) | Color subcarrier (MHz) | Sound carrier (MHz) |
|---|---|---|---|---|
| **VHF low-band channels** | | | | |
| 2 | 54–60 | 55.25 | 58.83 | 59.75 |
| 3 | 60–66 | 61.25 | 64.83 | 65.75 |
| 4 | 66–72 | 67.25 | 70.83 | 71.75 |
| 5 | 76–82 | 77.25 | 80.83 | 81.75 |
| 6 | 82–88 | 83.25 | 86.83 | 87.75 |
| FM | 88–108 | — | — | — |
| **VHF high-band channels** | | | | |
| 7 | 174–180 | 175.25 | 178.83 | 179.75 |
| 8 | 180–186 | 181.25 | 184.83 | 185.75 |
| 9 | 186–192 | 187.25 | 190.83 | 191.75 |
| 10 | 192–198 | 193.25 | 196.83 | 197.75 |
| 11 | 198–204 | 199.25 | 202.83 | 203.75 |
| 12 | 204–210 | 205.25 | 208.83 | 209.75 |
| 13 | 210–216 | 211.25 | 214.83 | 215.75 |
| **UHF channels** | | | | |
| 14 | 470–476 | 471.25 | 474.83 | 475.75 |
| 15 | 476–482 | 477.25 | 480.83 | 481.75 |
| 16 | 482–488 | 483.25 | 486.83 | 487.75 |
| 17 | 488–494 | 489.25 | 492.83 | 493.75 |
| 18 | 494–500 | 495.25 | 498.83 | 499.75 |
| 19 | 500–506 | 501.25 | 504.83 | 505.75 |
| 20 | 506–512 | 507.25 | 510.83 | 511.75 |
| 21 | 512–518 | 513.25 | 516.83 | 517.75 |
| 22 | 518–524 | 519.25 | 522.83 | 523.75 |
| 23 | 524–530 | 525.25 | 528.83 | 529.75 |
| 24 | 530–536 | 531.25 | 534.83 | 535.75 |
| 25 | 536–542 | 537.25 | 540.83 | 541.75 |
| 26 | 542–548 | 543.25 | 546.83 | 547.75 |
| 27 | 548–554 | 549.25 | 552.83 | 553.75 |
| 28 | 554–560 | 555.25 | 558.83 | 559.75 |
| 29 | 560–566 | 561.25 | 564.83 | 565.75 |
| 30 | 566–572 | 567.25 | 570.83 | 571.75 |
| 31 | 572–578 | 573.25 | 576.83 | 577.75 |
| 32 | 578–584 | 579.25 | 582.83 | 583.75 |
| 33 | 584–590 | 585.25 | 588.83 | 589.75 |
| 34 | 590–596 | 591.25 | 594.83 | 595.75 |
| 35 | 596–602 | 597.25 | 600.83 | 601.75 |
| 36 | 602–608 | 603.25 | 606.83 | 607.75 |
| 37 | 608–614 | 609.25 | 612.83 | 613.75 |
| 38 | 614–620 | 615.25 | 618.83 | 619.75 |
| 39 | 620–626 | 621.25 | 624.83 | 625.75 |
| 40 | 626–632 | 627.25 | 630.83 | 631.75 |

| Channel no. | Frequency range (MHz) | Picture carrier (MHz) | Color subcarrier (MHz) | Sound carrier (MHz) |
|---|---|---|---|---|
| 41 | 632–638 | 633.25 | 636.83 | 637.75 |
| 42 | 638–644 | 639.25 | 642.83 | 643.75 |
| 43 | 644–650 | 645.25 | 648.83 | 649.75 |
| 44 | 650–656 | 651.25 | 654.83 | 655.75 |
| 45 | 656–662 | 657.25 | 660.83 | 661.75 |
| 46 | 662–668 | 663.25 | 666.83 | 667.75 |
| 47 | 668–674 | 669.25 | 672.83 | 673.75 |
| 48 | 674–680 | 675.25 | 678.83 | 679.75 |
| 49 | 680–686 | 681.25 | 684.83 | 685.75 |
| 50 | 686–692 | 687.25 | 690.83 | 691.75 |
| 51 | 692–698 | 693.25 | 696.83 | 697.75 |
| 52 | 698–704 | 699.25 | 702.83 | 703.75 |
| 53 | 704–710 | 705.25 | 708.83 | 709.75 |
| 54 | 710–716 | 711.25 | 714.83 | 715.75 |
| 55 | 716–722 | 717.25 | 720.83 | 721.75 |
| 56 | 722–728 | 723.25 | 726.83 | 727.75 |
| 57 | 728–734 | 729.25 | 732.83 | 733.75 |
| 58 | 734–740 | 735.25 | 738.83 | 739.75 |
| 59 | 740–746 | 741.25 | 744.83 | 745.75 |
| 60 | 746–752 | 747.25 | 750.83 | 751.75 |
| 61 | 752–758 | 753.25 | 756.83 | 757.75 |
| 62 | 758–764 | 759.25 | 762.83 | 763.75 |
| 63 | 764–770 | 765.25 | 768.83 | 769.75 |
| 64 | 770–776 | 771.25 | 774.83 | 775.75 |
| 65 | 776–782 | 777.25 | 780.83 | 781.75 |
| 66 | 782–788 | 783.25 | 786.83 | 787.75 |
| 67 | 788–794 | 789.25 | 792.83 | 793.75 |
| 68 | 794–800 | 795.25 | 798.83 | 799.75 |
| 69 | 800–806 | 801.25 | 804.83 | 805.75 |
| 70 | 806–812 | 807.25 | 810.83 | 811.75 |

**UHF translator channels**

| Channel no. | Frequency range (MHz) | Picture carrier (MHz) | Color subcarrier (MHz) | Sound carrier (MHz) |
|---|---|---|---|---|
| 71 | 812–818 | 813.25 | 816.83 | 817.75 |
| 72 | 818–824 | 819.25 | 822.83 | 823.75 |
| 73 | 824–830 | 825.25 | 828.83 | 829.75 |
| 74 | 830–836 | 831.25 | 834.83 | 835.75 |
| 75 | 836–842 | 837.25 | 840.83 | 841.75 |
| 76 | 842–848 | 843.25 | 846.83 | 847.75 |
| 77 | 848–854 | 849.25 | 852.83 | 853.75 |
| 78 | 854–860 | 855.25 | 858.83 | 859.75 |
| 79 | 860–866 | 861.25 | 864.83 | 865.75 |
| 80 | 866–872 | 867.25 | 870.83 | 871.75 |
| 81 | 872–878 | 873.25 | 876.83 | 877.75 |
| 82 | 878–884 | 879.25 | 882.83 | 883.75 |
| 83 | 884–890 | 885.25 | 888.83 | 889.75 |

*Television frequency spectrum*

## RCI sub channels

| | | | | |
|---|---|---|---|---|
| 01 | 11–17 | 15.75 | 12.17 | 11.25 |
| 02 | 17–23 | 21.75 | 18.17 | 17.25 |
| 03 | 23–29 | 27.75 | 24.17 | 23.25 |
| 04 | 29–35 | 33.75 | 30.17 | 29.25 |
| 05 | 35–41 | 39.75 | 36.17 | 35.25 |

## CATV midband and superband channels
### Midband channels

| | | | | |
|---|---|---|---|---|
| A | 120–126 | 121.25 | 124.83 | 125.75 |
| B | 126–132 | 127.25 | 130.83 | 131.75 |
| C | 132–138 | 133.25 | 136.83 | 137.75 |
| D | 138–144 | 139.25 | 142.83 | 143.75 |
| E | 144–150 | 145.25 | 148.83 | 149.75 |
| F | 150–156 | 151.25 | 154.83 | 155.75 |
| G | 156–162 | 157.25 | 160.83 | 161.75 |
| H | 162–168 | 163.25 | 166.83 | 167.75 |
| I | 168–174 | 169.25 | 172.83 | 173.75 |

### Superband channels

| | | | | |
|---|---|---|---|---|
| J | 216–222 | 217.25 | 220.83 | 221.75 |
| K | 222–228 | 223.25 | 226.83 | 227.75 |
| L | 228–234 | 229.25 | 232.83 | 233.75 |
| M | 234–240 | 235.25 | 238.83 | 239.75 |
| N | 240–246 | 241.25 | 244.83 | 245.75 |
| O | 246–252 | 247.25 | 250.83 | 251.75 |
| P | 252–258 | 253.25 | 256.83 | 257.75 |
| Q | 258–264 | 259.25 | 262.83 | 263.75 |
| R | 264–270 | 265.25 | 268.83 | 269.75 |
| S | 270–276 | 271.25 | 274.83 | 275.75 |
| T | 276–282 | 277.25 | 280.83 | 281.75 |
| U | 282–288 | 283.25 | 286.83 | 287.75 |
| V | 288–294 | 289.25 | 292.83 | 293.75 |
| W | 294–300 | 295.25 | 298.83 | 299.75 |

# Who makes what

NOT ALL VCR BRANDS ARE MADE BY THE COMPANY ON THE nameplate. This appendix lists VCR brands and the companies that make the machines. The appendix is divided into three parts:

☐ Brand comparison—A comparison of major brands and their *OEM*s (original equipment manufacturers). The OEM is the company that made the VCR; the brand is the name on the name plate of the VCR.
☐ FCC ID comparison—A comparison of FCC identification numbers (found on the back of the VCR) and the original equipment manufacturer.
☐ UL listing ID comparison—A comparison of UL listing identification numbers (found on the back of the VCR) and the original equipment manufacturer.

## Brand comparison

The following lists major brands and their OEMs.

| Brand | Original equipment manufacturer (OEM) |
| --- | --- |
| Akai | Akai (first models) |
| | Mitsubishi |
| Canon | Matsushita |
| | National/Panasonic |
| Circuit City | Shintom |
| Colt | Shintom |
| Curtis Mathes | Matsushita |
| Daytron | Daewoo |
| Dynatech | Funai |
| Emerson | Mitsubishi |
| | Emerson |
| | Goldstar |
| Fisher | Fisher |
| Funai | Funai |
| General Electric | Hitachi |
| | Matsushita |

| Brand | Original equipment manufacturer (OEM) |
|---|---|
| Goldstar | Goldstar |
| Harmon Kardon | Mitsubishi |
| | NEC |
| Hitachi | Hitachi |
| Instant Replay | Panasonic |
| | Hitachi |
| JC Penney | Hitachi |
| | Matsushita |
| | JVC |
| Jensen | JVC |
| JVC | JVC |
| KLH | Shintom |
| KMC (K-Mart) | Sharp |
| Kenwood | JVC |
| | NEC |
| Kodak | Panasonic |
| Lloyds | NEC |
| | Funai |
| Logik | Shintom |
| Magnavox | Matsushita |
| Memorex | Goldstar |
| Midland | Samsung |
| Minolta | Hitachi |
| Mitsubishi | Mitsubishi |
| Montgomery Ward | Sharp |
| | Matsushita |
| Multitech | NEC |
| | Funai |
| National | Matsushita |
| NEC | JVC |
| | NEC |
| Olympus | Matsushita |
| Panasonic | Matsushita |
| Pentax | Hitachi |
| Philco | Matsushita |
| Pioneer | Sony |
| Portland | Daewoo |
| Quasar | Matsushita |
| Radio Shack/Realistic | Hitachi |
| | Sharp |
| | Toyomenka |
| RCA | Matsushita |
| | Hitachi |
| | Samsung |
| Sampo | Sampo |
| Samsung | Samsung |
| Sansui | Sansui |

| | |
|---|---|
| Sanyo | Sanyo |
| Scott | Emerson |
| Sears | Hitachi |
| | Fisher |
| | Sanyo (Beta) |
| Sharp | Sharp |
| | Matsushita (camcorder) |
| Singer | Shintom |
| Sony | Sony |
| | Aiwa (SL5000 & SL5200 Beta) |
| Sound Design | Funai |
| Supra | Akai |
| | Samsung |
| | Funai |
| | Supra |
| Sylvania | Matsushita |
| Symphonic | Funai |
| | NEC |
| Teac | JVC |
| | Funai |
| Teknika | Matsushita |
| | Teknika |
| Technics | Matsushita |
| Toshiba | Toshiba |
| | Sony |
| Totevision | Goldstar |
| | Samsung |
| | Daewoo |
| Toyomenka (TMK) | Funai |
| Unitech | Samsung |
| Vector Research | NEC |
| | Funai |
| Video Concepts | Mitsubishi |
| Zenith | JVC (VHS) |
| | Sony (Beta) |

# FCC number comparison

VCRs must be registered with the Federal Communications Commission (FCC) and undergo rigorous testing to ensure that they meet various minimum standards of operation (the quality of the VCR is not at issue; the FCC is concerned with the amount of radio frequency interference produced by the VCR). The FCC ID number is printed on the back of the VCR. The prefix of the number can be used to help identify the original manufacturer of the VCR.

| FCC ID Number | Manufacturer |
|---|---|
| A3D | NEC |
| A3L | Samsung |
| A7R | Orion |
| AAO | Radio Shack |
| ABA | Hewlett Packard |
| ABL | Hitachi |
| ABW | JC Penney |
| ABY | Motorola |
| ABZ | Motorola |
| ACA | Yorx Electronics |
| ACB | Phonotronics |
| ACJ | Matsushita |
| ADT | Funai |
| AES | Uniden |
| AEZ | Sanyo |
| AFA | Fisher |
| AFL | Sharp |
| AFR | Curtis Mathes |
| AGH | Koshin Denki Kogyo (KDK) |
| AGI | Toshiba |
| AGV | Montgomery Ward |
| AHA | RCA |
| AHK | Matsushita (Panasonic) |
| AIX | Sylvania |
| AJT | Harmon Industries |
| AJU | GE |
| AJX | Toshiba America Consumer Products |
| AK8 | Sony |
| ALE | Samsung |
| AKJ | GE |
| AKL | Act Electronics |
| AKN | Toshiba (Sears) |
| AKO | Toshiba (JC Penney) |
| AKP | Toshiba (Radio Shack) |
| AKQ | Toshiba (Wards) |
| AKR | Sears Roebuck |
| AKX | Shinwa |
| AKZ | Sanyo (Sears) |
| ARS | AOC |
| ASH | Akai |
| ASI | JVC |
| ATA | Sharp |
| ATH | E F Johnson |
| ATO | Zenith Electronics |
| ATP | Advent |
| ATQ | Philips Hi Fi Products |

| | |
|---|---|
| BAA | Taiyo Musen (TMC) |
| BAB | Fujitsu Ten |
| BCF | General Instrument |
| BEJ | Goldstar |
| BEK | Nady |
| bgb | Mitsubishi |
| BOU | Philips |
| C5F | Daewoo |
| EEK | Ampex |
| EEN | Philips (Netherlands Division) |
| EEV | Sanyo |
| E0Z | Shintom |
| EW7 | Video Technology Engineering (Vtech, Tropez) |
| FAG | Nokia |

## UL listing comparison

The UL listing denotes the VCR meets or exceeds minimum safety requirements as an electric appliance. The UL listing number is printed on the back of the VCR. The prefix of the number can be used to help identify the original manufacturer of the VCR.

| Number | Manufacturer | Brand Names |
|---|---|---|
| 16M4 | Samsung | Supra, Multitech, Unitech, Totevision, Cybrex, GE, RCA, Dumont |
| 174Y | Toshiba | Sears |
| 216V | Toshiba | |
| 238Z | Hitachi | RCA, GE, JC Penney, Pentax |
| 246T | Daewoo | Portland |
| 289X | Toyomenka | TMK |
| 333Z | Symphonic | Teac, KTO, Realistic, Multitech, Funai |
| 403Y | Fisher/Sanyo | Realistic, Sears |
| 41K4 | Daewoo | |
| 439F | Jvc | Zenith, Kenwood, Sansui |
| 444H | Zenith | Zenith, Memorex, Curtis Mathes |
| 44L6 | TMC/Orion | Emerson, Lloyds, Broksonic, Wards, KMC |
| 463K | Nikko | |
| 504F | Sharp | Wards, KMC |
| 51K8 | Portavideo | |
| 536Y | Mitsubishi | Emerson, Video Concepts, MGA |
| 570F | Sony | Zenith |
| 628E | Samsung | Curtis Mathes |
| 637J | Yamaha | |
| 679F | Panasonic | RCA, GE, Magnavox, Quasar, Canon, Philco |
| 781Y | NEC | Dumont, Vid Concepts Vector, Sears |
| 828B | Matsushita | Quasar, Panasonic |

| Number | Manufacturer | Brand Names |
|--------|-------------|-------------|
| 86B0 | Goldstar | Realistic, JC Penney, ToteVision, Shintom, Sears, Memorex |
| 8L55 | Vtech | Tropez |
| 932T | AOC | AOC |
| 91B9 | Pioneer | |

# Glossary

**A/B switch**   A two-way switch that is used to alternately divert a signal one of two ways. Also can be used to select from either of two incoming signals.

**ac**   Abbreviation for alternating current. Current that fluctuates to positive and negative values about a zero point. The current available at wall outlets.

**AFC**   Abbreviation for automatic frequency control, a circuit built into some VCRs and TVs to lock on an incoming channel automatically.

**AGC**   Abbreviation for automatic gain control. In a TV or VCR, AGC is a circuit that automatically adjusts the incoming signal to the proper levels for display or recording. In a video camera, AGC is a circuit that automatically adjusts the sensitivity of the pickup tube to render the most pleasing image.

**attenuator**   A passive (that is, not powered by electricity) device used to reduce the power of a signal.

**azimuth**   The tilt of a magnetic head or recorded track with respect to the direction of travel of the tape.

**balance**   Equal signal strength provided to both left and right stereo output channels. Balance can be adjusted to provide more signal strength to one channel than the other.

**balun**   See *matching transformer*.

**baseband**   Separate audio and video signals from a VCR or similar signal source.

**Beta**   A VCR format pioneered by Sony in the mid 1970s.

**Beta I/II/III**   The three tape record/playback speeds used in Beta decks. Beta I speed is fastest; Beta III is slowest.

**bit**  A binary number: 1 or 0. A certain number of bits makes a byte. In most computer applications, eight bits make a byte.

**burst**  A component of a television signal that carries color information.

**C band**  The frequency band used for most commercial satellite television transmissions.

**CATV**  An acronym for community antenna television, now generally regarded as *cable TV*.

**CCD**  Short for charge-coupled device, a type of light-sensitive integrated circuit that is used as an imaging device in video cameras.

**CCTV**  Acronym for closed-circuit television. The term generally applies to cable TV.

**CRT**  Short for cathode-ray tube, the screen in a TV.

**CTL**  An abbreviation for control, having two meanings: (1) The indexing/address system used on the latest models of VHS videocassette recorders. (2) In reference to the control track on a tape or a control head in a VCR.

**cable converter**  An electronic device used to tune to channels broadcast through a cable system. A converter is different than a cable decoder, which unscrambles an encrypted video/audio signal.

**camcorder**  A combination video camera and VCR.

**capstan**  A motor-driven shaft that pulls the videotape through the transport mechanism. The capstan is supplemented by the pinchroller.

**carrier**  (1) A radio-frequency signal that can have another signal superimposed on it. (2) The mechanism that holds the cassette inside the VCR; part of the cassette-lift mechanism.

**cassette**  A cartridge that holds videotape. Also called the shell.

**cassette-in switch**  A small leaf switch used to detect when a cassette tape is fully loaded in the VCR.

**cassette-lift mechanism**  The mechanism that brings the cassette into threading position. The cassette-lift mechanism includes the carrier.

**chroma**  Short for chrominance. *Chroma* is the color component of the video signal.

**clipping**   An effect of distortion where the peaks of driven signals are chopped off. Clipping usually occurs in the amplifier when it is turned up too high, but it also can occur in maladjusted circuits in a VCR or TV set.

**clock**   (1) An electronic "metronome" used for timing signals in a VCR's digital circuits. (2) The timing of a sequence of signals.

**coax**   Short for coaxial, a type of rounded cable made with a solid center conductor that's completely encased in a plastic covering. Around the plastic is a jacket of braided shielding wire and sometimes an aluminum sheath. It is the most popular type of cabling used for connecting video components.

**color burst**   The signal, at approximately 3.57 MHz (megahertz) in the video bandwidth, that stores the instantaneous intensity and hue of the color for a particular spot in the TV image.

**color-under**   A method of placing the color (or chroma) information below the luminance information in the video bandwidth. The color-under (or heterodyne) method is used to enable accurate recording and playback of the video image.

**comb filter**   An electric filtering system designed to pass a certain set of frequencies but reject others.

**composite video**   A picture signal combined with synchronization and (possibly) color information. Usually called baseband video, or just video.

**common**   The ground point of a circuit or a common path for an electrical signal.

**control head**   The magnetic head in a VCR that records and plays back the control track.

**control track**   A linear track, consisting of 30- or 60-Hz pulses, placed on the bottom of videotape that aids in proper playback of the video signal.

**crosstalk**   (1) A signal from one stereo channel that bleeds into the other. (2) A signal from a video track on a tape bleeding into the signal on the adjacent track.

**cue**   (1) To view a tape in the forward direction in faster-than-normal speed. (2) A special signal or track recorded on the tape with the picture and sound.

**dc** Abbreviation for direct current. It is current such as that from a battery where the voltage level remains the same (either positive or negative) in respect to ground.

**decibel (dB)** A unit of power measurement. A 3 dB rise in signal strength represents a 100% increase (or doubling) in power.

**demodulate** To remove the carrier signal and leave only baseband audio and video.

**depth-multiplex recording** The process of recording high-fidelity audio on VHS hi-fi VCRs. The system uses separate rotating heads to record the audio portion.

**dew sensor** A passive electrical device used in most VCRs that detects the presence of moisture. Most dew sensors measure resistance; the resistance drops when moisture is present.

**digital** A signal composed of two signal states—on or off (often expressed as 1s and 0s). The alternate to digital is analog, where the signal is continuously variable.

**Dolby** The trade name for a popular audio noise-reduction system.

**down-converter** An electronic device that converts very high frequency satellite signal to a more manageable frequency, often 70 MHz or a block of frequencies between 430 to 930 MHz.

**dropout** Missing information. In video, dropout is usually caused by missing oxide emulsion on the tape.

**dub** To copy a tape.

**duty cycle** The amount of time a signal is turned on. The longer the signal is on, the longer the duty cycle.

**dynamic range** The range of volume from softest to loudest sounds (the actual sound levels themselves are not a consideration, but it usually starts at "no sound" and goes up from there). Expressed in dB (decibels). The higher the dB, the wider the dynamic range. Also used to denote picture brightness range.

**E-E** Abbreviation for electronics-to-electronics, a signal that passes through the processing circuits of a VCR directly to a television set or monitor as opposed to viewing a recorded signal.

**EP** Abbreviation for extended play, the slowest speed on a VHS VCR. On earlier models, EP was called SLP, probably for "super long play." See also *SP* and *LP*.

**end-of-tape sensor**   A sensor system used in all VCRs to detect the start or end of the videotape.

**F connector**   The standard connector used with coaxial cable and the RF inputs/outputs of most video equipment.

**feedhorn**   In satellite television systems, a signal collector placed in front of a low-noise amplifier.

**field**   One-half of a video field, comprising the odd or even scan lines. There are 60 fields in one second of video.

**filter**   An electrical circuit designed to prevent the passage of certain frequencies.

**flagging**   Bending at the top of a picture played back by a VCR.

**footprint**   The signal coverage of a satellite.

**frame**   One complete video picture, comprising both odd and even fields. There are 30 video frames per second.

**frequency response**   As used in audio applications, sonic range; the highest and lowest audio frequencies that can be accurately reproduced. As used in video applications, brightness range; the darkest and lightest tones that can be accurately reproduced.

**gain**   Amplification, usually expressed in dB.

**ghost**   A duplication of the video image (a type of visual echo), usually caused by a reflection in the broadcast signal or an improperly terminated signal input or output.

**gigahertz (GHz)**   A measurement of frequency. One gigahertz is equal to one billion cycles per second.

**ground**   Refers to the point of (usually) zero voltage and can pertain to a power circuit or a signal circuit.

**guard band**   A blank area on the tape separating two signal tracks.

**HQ**   A set of circuits used in some VHS videocassette recorders that provides improved video resolution. The HQ "standard" is a set of four separate picture-improvement circuits (true HQ), although some VCRs only use one or two of the circuits (mock HQ).

**helical scan**   The technical name for the way the video heads in a VCR record and play back picture information. Also used to record and play back stereo hi-fi audio.

**hertz**   Abbreviated *Hz*. A unit of measurement used for expressing cycles per second, named after German physicist H.R. Hertz.

**365**

One hertz equals one cycle per second. Also used with the letters k, M, and G as multipliers to indicate thousands, millions, and billions, respectively.

**heterodyne**   See *color-under*.

**hi-fi**   A general term used to denote the capability of reasonably high-fidelity audio playback. Hi-fi VCRs are equipped with separate rotating audio heads (VHS), or they mix sound and picture in one signal and record both simultaneously on tape (Beta).

**hiss**   Audible high-frequency noise. Can be caused by the recording medium (such as videotape) or circuitry.

**hum**   A low-level, low-frequency electrical noise usually caused by the alternating current in household power lines interfering with audio and video signal circuits. Hum is often picked up in cables extending between the VCR and TV or hi-fi but also can be caused by an ungrounded circuit. The hum is audibly apparent as a low-frequency (60 Hz) tone; video "hum" is a dark bar that floats through the picture.

**IC**   Abbreviation for integrated circuit. Also called a chip. A complete electrical circuit housed in a self-contained package. See also *LSI*.

**IR**   Abbreviation for infrared. Used in VCRs for remote sensing and remote control applications.

**idler**   A wheel most often used to drive the supply and take-up tape reel spindles. The idler is powered by a separate motor (rarely) or driven by a belt connected to the capstan motor.

**impedance**   The degree of resistance that an alternating electrical current (ac) encounters when passing through a circuit, device, or wire. The amount of impedance is expressed in ohms ($\Omega$).

**impedance roller**   Metal or plastic rollers used in most VCRs to provide an even and steady flow of tape through the transport mechanism.

**insertion loss**   The attenuation or loss of power caused by the insertion of a component or circuit into the signal path.

**interlace**   The process of combining the odd and even lines of subsequent video fields to generate a complete frame.

**jack**   A term used for the cable connector in audio and video equipment.

**Ku band**   The 12 GHz frequency band used for some satellite television transmissions.

**Kelvin (degrees)**   A temperature measurement used in grading low-noise amplifiers, as used in satellite television systems.

**kilohertz (kHz)**   A measurement of frequency. One kilohertz is equal to one thousand cycles per second.

**LED**   Abbreviation for light-emitting diode, a unique type of semiconductor that is made to emit a bright beam of light. Often used as a panel indicator, but also is used in remote sensing systems and infrared remote control devices.

**LNB**   Abbreviation for low-noise block down-converter. In satellite TV systems, the part that receives the signal and amplifies it for use by the receiver. An LNB contains both a low-noise amplifier and block down-converter.

**LP**   Long play, the intermediate recording speed found on many (but not all) VHS VCRs. See also *SP* and *EP*.

**LSI**   Abbreviation for large-scale integration. A complex integrated circuit (IC) that is a combination of many ICs and other electronic components that are normally packaged separately. LSI (and *VLSI*, for very large scale integration) chips are used in the latest video equipment and held proprietary by the manufacturer.

**367**

**loading**   A function of the VCR where the cassette is drawn into the machine and made ready for threading. Loading is automatic in front-loading VCRs and manual for top-loading VCRs.

**logic**   Primarily used to indicate digital circuits or components that accept one or more signals and act on those signals in a predefined, orderly fashion.

**luminance**   A term used to denote the brightness or black-and-white picture of a video image.

**LV**   Abbreviation for LaserVision, the trade name of optical video discs.

**MOS**   Abbreviation for metal-oxide semiconductor, a type of imaging device used in certain video cameras and camcorders.

**MTS**   Abbreviation for multichannel television sound, the stereo broadcasting standard used in the U.S.

**M load**   The tape threading pattern used by 8mm and VHS VCRs. Also see *U load*. The term is derived by the M shape of the tape when threaded in the VCR transport mechanism.

**matching transformer** A device used to match the impedance between one cable or device to another cable or device. Matching transformers are most often used to balance the impedance of an outdoor antenna (usually 300 Ω) to the impedance of modern TV sets and VCRs (usually 75 Ω).

**megahertz (MHz)** A measurement of frequency. One megahertz is equal to one million cycles per second.

**micron** A unit of linear measurement; one micron or 1 μm (micrometer) is equal to 0.001 millimeter.

**microprocessor** A special integrated circuit that performs semi-intelligent functions based on instructions written in a program. Fundamentally, a microprocessor is a hardware device that can be electrically "rewired" by using new software. The microprocessor is often considered the brain of a computer or circuit. Many of the later VCRs use microprocessors to control system functions.

**midband** The frequencies found between channels 6 and 7, designated channel A through I. Often used in cable television systems.

**modulation** A way in which one signal modifies or controls another signal for such purposes as enabling it to carry information. Often used to describe radio frequency (RF) transmission. *FM* is frequency modulation; *AM* is amplitude modulation.

**monitor** A video display. A monitor is like a TV except it lacks the ability to tune in channels. A monitor might or might not have a sound amplifier and speaker.

**monitor/TV** A combination monitor and television set.

**noise** An unwanted signal. Audio noise is usually heard as hiss; video noise is usually seen as "salt and pepper" flecks.

**NTSC** National Television Standards Committee. A group of businesses and engineers originally created to decide on early standards for color and black-and-white television in the U.S. The NTSC system is also used in Japan. Other television standards around the world include PAL (most of Europe) and SECAM (France, parts of Africa, and the former Soviet Union).

**ohm (Ω)** The unit of measure of impedance or resistance.

**oxide** The magnetic coating applied to videotape that actually stores the video and audio signals generated by the VCR.

**PCM**   Abbreviation for pulse code modulation, a way of digitally recording an audio signal. Used in most 8mm tabletop decks and in some high-end VHS recorders.

**phone plug**   A certain type of cable connector often used for headphone and microphone audio applications. It has a single stem, with parts of the stem insulated to provide for additional contact(s). Phone plugs come in many sizes: ¼-inch phone plugs are large and used for headphones and some microphones; ⅛-inch (or miniature) phone plugs are smaller and used mainly for lightweight headsets. See also *phono plug*.

**phono plug**   A certain type of cable connector used in home audio and video applications. It is composed of a round center stem and an insulated metal cap. Also called an RCA plug. See also *phone plug*.

**pickup tube**   The photoreceptor of a video camera.

**pinchroller**   A rubber roller that presses against the capstan to pull the tape through the transport mechanism.

**RAM**   Abbreviation for random-access memory. As used in VCRs, electronic memory used to store programming or picture information temporarily.

**RF**   Abbreviation for radio frequency. In television, VHF, UHF, and cable TV signals are considered RF.

**RF modulator**   A device that adds a carrier to baseband audio and video so the signal can be picked up by a VCR or TV.

**resistance**   Opposition to direct electrical current (dc), expressed in ohms ($\Omega$).

**resolution**   The clarity or sharpness of the picture. Resolution is most often stated in the number of total lines that make up an image or in megahertz (millions of cycles per second).

**review**   To view a tape in the reverse direction in faster-than-normal speed.

**roller**   A rubber tire or wheel.

**SAP**   Abbreviation for special audio program, a separate sound signal used in some multichannel television sound (MTS) broadcasts.

**SP**   Abbreviation for short play (or standard play), the fastest speed setting in VHS VCRs. See also *LP* and *EP*.

369

**scanner**   The video-head drum assembly.

**search**   To view a tape in forward or reverse in faster-than-normal speed.

**separation**   The complete electrical separation of two or more signals that usually refers to either right and left stereo channels, audio and video signals, or luminance and chrominance signals.

**servo**   An electronic circuit that modifies its output in accordance with a constantly varying input signal.

**shell**   A cassette tape cartridge.

**shielding**   A conductive material surrounding, but insulated from, an electrical conductor.

**signal**   The desired portion of electrical information.

**signal-to-noise ratio**   Abbreviated *S/N*. The ratio of wanted signal to unwanted signal, usually expressed in dB. The higher the number, the better.

**splitter**   A device that separates a signal into two (or more) signal paths of substantially equal strengths.

**stereo adaptable**   A VCR or TV that can be made to receive MTS stereo audio broadcasts by the addition of an external decoder.

**stereo ready**   A VCR or TV that already has the necessary circuitry to decode MTS stereo audio broadcasts.

**subcarrier**   A frequency-derived signal of a main signal.

**Super VHS**   A VHS format that enables recording and playback of very high resolution video.

**supply-reel spindle**   The left spindle in the VCR that drives the supply tape reel of a cassette.

**Surround**   A trade name for a stereo audio enhancement scheme that places extra speakers in the center and rear of the room.

**sync**   Short for synchronization, a broad term to indicate the proper order of electrical signals to generate and display sound and picture from a videotape.

**take-up reel spindle**   The right spindle in the VCR that drives the take-up tape reel of a cassette.

**tape-guide spindle**   One of several posts used in VHS and Beta VCRs to position the tape accurately in the transport mechanism.

**tape-load switch**   A small leaf switch used in some VCRs to detect when the threading mechanism has fully threaded the tape around the video heads.

**tape-tension lever**   A lever that is used to detect tape supply-reel tension. The lever is connected to a brake; when the tension decreases, the brake tightens to slow the reel. The reverse occurs when the tape tension increases.

**terminating resistor**   A resistor (usually 75 $\Omega$) attached to the end of a cable or to an input or output on a piece of video equipment. The resistor restores proper system impedance.

**threading**   A function of the VCR where tape is withdrawn from the cassette and placed against the rollers and magnetic heads to facilitate recording and playback. Threading is accomplished by the threading mechanism.

**threading ring**   The threading mechanism used in Beta VCRs.

**torque**   The amount of mechanical force exerted on an object. In VCRs, torque usually relates to the rotational force exerted by the supply and take-up reels.

**track**   (1) One of several discrete signals applied to videotape during recording. (2) Parallel bands of video information recorded in a diagonal on the videotape.

**371**

**tracking**   The alignment of the video heads in a VCR to tracks already recorded on the tape.

**tracking control**   An electronic control that varies the tracking timing of the VCR.

**transponder**   A channel on a satellite.

**trap**   A filter.

**twinlead**   TV hookup cable that consists of two wires separated a certain distance by a plastic spine.

**U load**   The tape-threading pattern used by Beta VCRs. Also see M load. The term is derived by the U shape of the tape when threaded in the VCR transport mechanism.

**UHF**   Abbreviation for ultra high frequency. TV channels 14 through 83, occupying the frequency spectrum of 300 to 3,000 MHz.

**VCR**   Abbreviation for videocassette recorder.

**VCP**   Abbreviation for videocassette player.

**VHF**   Abbreviation for very high frequency, TV channels 2 through 13, occupying the frequency spectrum of 30 to 300 MHz.

**VHS**   Abbreviation for video home system, a VCR format pioneered by JVC.

**video-head drum**   The cylindrical-shaped mechanism that houses the video heads (it also might house the hi-fi audio heads in a VHS hi-fi VCR).

**writing speed**   The effective video head-to-tape speed during playback and recording. The writing speed considers the rotational speed of the video heads, not just the linear speed of the tape as it travels through the transport mechanism of the VCR.

# Index

## 8

8mm, 3, **4**, 48
  hi-band, 68
  video head, **70**
  video tracks recorded on, **69**

## A

A/B switch, 82, **83**, **200**
ac plug, **86**
amplified coupler, **84**
amplifiers, signal, 84, **84**
attenuators, 82-83
audio (*see also* sound)
  baseband, 30
  hi-fi, 55-58
  PCM, 69
  signal recording/playback, 46
  weak, 314
AUDIO DUB, 71
audio dubbing, 16-17, **17**
audio heads, 7, **10**
  adjusting, 244, **244**, **245**
  rotating, 11
  stereo, 9-10
AUDIO IN, 72
AUDIO OUT, 72
audio tapes, **4**
automatic color control (ACC), 45
automatic phase control (APC), 45
automatic track finding (ATF), 48
autotracking, 68
azimuth, 51

## B

Baird Televisor, 24
baluns, 81, **81**
band splitter, 82
bandpass filter, 44
bar-code programming, 20

baseband audio, 30
baseband signal, 29-30
baseband video, 30, 93-95
batteries
  camcorder, 326-327
  leakage from, 193-194
  problems with, 318-319
belts, **147**
  maintenance, 146-149
Beta, 2, 47-48
  cover release pin, **172**
  hi-fi audio, 55-57
  hi-fi audio signal layouts, **56**
  misthreading, 314-315
  playback circuit, **47**
  recording circuit, **47**
  Super, 18-19, 59
  videocassette, **3**
books, 338-339
brakes, **150**, **276**
  maintenance, 149
brightness noise reduction, 18
burst, 27, 66

## C

cable-ready VCRs, 12-13
cable box control, 70
cables, 77-78, 101-103, 211
  A/B switch, 82, **83**, **200**
  attentuators, 82-83
  band splitter, 82
  coaxial, 78, **80**, 101-103, 106-107, **138**
  connecting, 79
  connectors and, 103-106
  filters, 83-84
  maintenance, 137-138
  matched impedances, 78-79
  matching transformers, 81, **81**
  ribbon, **185**
  shielded, 101-103

Illustrations are in **boldface**.

cables, *continued*
    signal amplifiers, 84
    signal splitter, 81-82
    twin lead, 79
    two-way signal splitter, **82**
    types of, 79-80
camcorders, 3, 5-6, **5**, 321-328
    batteries, 326-327
    care and maintenance, 325-326
    cleaning
        heads, 322-323
        lens, 324
        parts, 323-324
    disassembly, 321-322
    dusting, 323
    preventive maintenance, 324
    replacing parts, 323-324
    tips when traveling with, 327-328
    UV filters, 326
capacitance, 102, 105
capstan, 37-40, **39**
    shaft bearings, 278, **278**
capstan pinchroller, 37, **39**
    adjusting, 246
    maintenance, 146-149
    rejuvenator lathe, 129-130, **130**, **148**, **149**
cassettes (*see* tapes)
cathode ray tube (CRT), 24
CHANNEL, 71
chrominance, 27
    converting for playback, 45-46
    converting for recording, 44
cleaner/degreaser, 126
cleaning, 133-181
    battery acid, 194
    camcorders, 323-324
        heads, 322-323
        lens, 324
    electrical contacts, 151-152
    end-of-tape sensor, 162-163, **163**
    exterior dust and dirt, 136-137
    front-panel controls, 165-166, **165**
    input/output terminals, **192**
    interior, 149-151
    personal safety, 135-136
    preventive maintenance checkup, 133-135
    printed circuit board, 150
    reel spindles, 149, 163-165, **270**
    remote control, 167-171, **191**
    sand/dirt/dust, 191-192
    soldering iron, 344
    sticky/staining liquid from VCR, 190

    supplies for, 125-128
    tires/pinchrollers/belts, 148
    tuner, 161-162
    video heads, 152-159, **153**, **155**, **156**
CLOCK, 71
co-channel interference, 205, 210
coaxial cable, 78, 101-103, **138**
    choosing, 106-107
    cutaway view of, **80**
color burst signal, 66
color noise reduction, 18
color processors, 96-97
color subcarrier signal, 27
composite video, 29
connectors, 71-72, 103-106
    cleaning, 106
    F, **80**, 345-347, **346**
    gold-plated, 105
    nickel-plated, 106
    phono, **80**
    quality, 105-106
    Y-C, 61
continuous wave (CW), 44
control-track head, 9, 38, 47
controls, 13-15, 70-71, 211
    cleaning, 165-166, **165**
    indicators not functioning, 288-290
    jog/shuttle, 166
    VCR not responding to, 285-288
Copyguard, 217-218
counter
    elapsed-index, 67
    real-time, 67
coupler, amplified, **84**
cross-modulation, 205, 210
cross-channel interference, **51**
cue, 16, 70


**D**

decoders, 98
demagnetizing, 162
depth multiplex recording, 58
detail enhancer, 18
dew sensor, 257, **257**
dielectric, 137
digital effects, 19, 62-64
dock, 31
Dolby, 10, 46
    pro-logic surround, 108
dropout, 318
dropout compensator (DOC), 45

dubbing, audio/video, 16-17, **17**
duty cycle, 39

## E

EJECT, 32, 70
elapsed-index counter, 67
electrical shock, 307-310
electrostatic discharge, 140, 344
equalization, 44
events, 13
extended play (EP), 3

## F

F connector, **80**, 345-347, **346**
Farnsworth, Philo T., 24
FAST FORWARD, 13, 14, 15, 70
   won't operate, 279-281
fast scan, 16, **52**
Federal Communications Commission
   (FCC), ID numbers, 357-359
filters, 83-84, 207-208
  FM, 83
  highpass, 83
  UV, 326
flagging, 317-318
flowcharts, 229-231 (*see also* servicing; troubleshooting)
  cassette will not load, 261
  cassette will not reject, 265
  fast forward won't operate, 279
  front-panel indicators not functioning, 288
  list of, 248-249
  remote control does not operate properly,
   303
  rewind won't operate, 279
  search does not work correctly, 282
  snowy video; poor audio, 300
  sound okay; video not okay, 295
  timer does not operate properly, 291
  tracking control has no effect, 284
  tuner channels don't change on TV, 293
  using, 247-248
  VCR does not respond to front-panel controls, 286
  VCR does not turn on, 250
  VCR eats tape, 274
  VCR makes unusual mechanical noises,
   306
  VCR overheats, 311
  VCR turns on but nothing else, 256
  VCR will not play tape, 271

VCR will not thread tape, 269
video okay; sound not okay, 297
you receive electrical shock when touching VCR, 308
flying erase heads, 11, 53-54
  adjusting, 245
  effects of insert editing recording, **54**
  position on video drum, **54**
flywheel, **39**
FM filter, 83
formats, 2-5, **7**
  8mm, 3, 48
  Beta, 2, **3**, 35-36, 47-48, 55-57, 314-315
  Super Beta, 18-19, 59
  Super VHS, 18, 59-62
  VHS, 2, **3**, 57-58
freeze frame, 15-16, 63
Freon, 126
frequency
  radio, 29-30
  ultra high, 12
  very high, 12
frequency meter, 119-120, **120**
frequency spectrum, 349-353
full-erase head, 6
fuse, 250-251, **251**
fuzzy logic, 68

## G

grease, 127-128, 159-161, **160**, **161**
grounding rod, 85

## H

half loading, 34
Hall-effect ICs, 37
head drum, 37-38
HEADPHONE, 72
heads, 6-11
  audio, 7, **10**, 244, **244**, **245**
  cleaning, 152-159
  cleaning camcorder, 322-323
  control-track, 9, 38, 47
  flying erase, 11, 53-54, **54**, 245
  full-erase, 6
  geometries of, **50**
  linear stereo, 10, **10**
  magnetic, 42, **42**
  rotating audio, 11
  self-cleaning video, 21
  stereo audio, 9-10
  switching, 54-55, **55**

*375*

heads, *continued*
    video, **7**, 8, **11**, 48-49, 52-53, 152-159
heterodyning, 45
hi-band 8mm, 68
hi-fi audio, 55-58
    Beta, 55-57
    Beta signal layouts, **56**
    head position, **57**
    limitations, 58
    VHS, 57-58
    VHS signal layout, **58**
high quality (HQ), 17-18
highpass filter, 83
home theater, 107-108
horizontal synchronization, 27, 44


**I**

iconosocope camera, 24
idler tire, **41**, **147**
idler wheel pivot, **276**
image enhancers, 96
impedance, 101-102
index, 19
infrared light detector, 120-123, **122**, **124**
    circuit description, 122-123
    using, 123
input/output terminals, 15, **192**
insertion loss, 95
installation, 76-77
    back panel diagram, **73**
    cables, 77-78
        A/B switch, 82, **83**, **200**
        attenuators, 82-83
        band splitter, 82
        coaxial, 78, 101-103
        connecting, 79
        filters, 83-84
        matched impedances, 78-79
        matching transformers, 81
        shielded, 101-103
        signal amplifiers, 84
        signal splitter, 81-82
        twin lead, 79
        two-way signal splitter, **82**
        types of, 79-80
    debugging, 86
    grounding rod, 85
    home theater system, 107-108
    hooking VCR to stereo receiver, 89-90, **89**
    hookup of cables, 77-86
    improper connection when watching
     Macrovision tapes, **199**

location/placement, 100-101
    remote control location, 77
    testing, 86-87
    tips, 85-86
    troubleshooting guide, 88
    ventilation, 76-77
interference, 202-211
    amateur radio, 204, 210
    attenuators, 208
    automobile ignition, 209
    broadcast sources, 203-205
    cable TV, 206-207
    CB, 204, 209
    channel 8 tweet, 211
    co-channel, 205, 210
    color oscillator, 210
    cross-channel, **51**
    cross-modulation, 205, 210
    dimmer switch noise, 209
    filters, 207-208
    fluorescent lamp noise, 208-209
    FM broadcast, 210
    from within TV set, 205-206
    horizontal, 211
    microwave oven, 209
    motor noise, 208
    nonbroadcast sources, 202-203
    problem/solution chart, 208-211
    remote control, 213-214
    traps, 207


**J**

jitter, 64-66
job/shuttle control, 19-20
    cleaning, 166


**K**

Kappell, Frederick R., 25


**L**

leakage current test, 186-187, **186**, **187**
light-emitting diode (LED), 120-122
    testing, 289, **289**
LINE, 71
linear stereo heads, 10, **10**
loading, 31-32
    front, **15**, 31-32
    half, 34
    M, 33
    tape, 14
    top, **14**, 32

logic probe, 114-117, **115**, **316**, **317**
    using, 115-117, **116**
logic pulser, 117, **117**
long play (LP), 3
luminance
    converting for playback, 45-46
    converting for recording, 43-44
luminance signal, 27

# M

M loading, 33
Macrovision, 97, 217-225
    "black box" buster, 224-225
    definition/description, 217-218
    history, 219-220
    identifying tapes with, 223-224
    improper VCR connection, **199**
    problems associated with presence of,
     221-222
    remedies/solutions, 222-223
    television problems and, 221
magazines, 337-338
magnetic heads, 42, **42**
maintenance (*see also* cleaning; servicing;
     troubleshooting)
    adjustments, 241-247
    brakes/torque limiters, 149
    cables, 137-138
    camcorders, 325-326
    checkout, 167
    demagnetizing, 162
    disassembling VCR, 139-140
    final inspection, 166-167
    general, 136-138
    log, 167, 178-181
    oiling/lubricating, 159-161, **160**, **161**
    preliminary inspection, 141-146
    preventive, 133-135, 139-152
    procedures, 241-247
    schedule, **134**
    soldering iron, 344
    tires/pinchrollers/belts, 146-149
    transmission system/clutch, 164-165
manuals, 136, 240-241
memory, random access, 63
memory effect, 327
MICROPHONE, 72
moiré, 53
mosaic, 63
MTS DECODER, 72
multichannel TV sound (MTS), 16, 98
multiplex (MPX), 16

# N

noise (*see also* interference)
    mechanical, 305-307
noise bars, 19
    video heads and, 51-52
noise reduction
    brightness, 18
    color, 18
    Dolby, 10, 46, 108
numeric address, 19

# O

oil, 127-128, 159-161, **160**, **161**
on-screen programming, 20
oscilloscope, 118-119
OTR, 71
oxidation, 104-105

# P

PAUSE, 14, 15, 19, 71
persistence of vision, 25
phono connectors, **80**
picture
    color processors, 96-97
    flagging, 317-318
    flickering, 198-199
    ghosts, 201, **201**
    image enhancers, 96
    obtaining better, 87-88
picture tubes, 27-28
picture-in-picture (PIP), 63, 64
PLAY, 13, 14, 70
playback
    audio signal, 46
    chrominance conversion for, 45-46
    control-track, 47
    luminance conversion for, 45-46
    VHS circuit, **46**
playback speeds, 3-4
PlusCode numbers, 20
posterization, 63
POWER, 70
power cords
    ac plug, **86**
    polarity, **87**
    testing with volt-ohmmeter, **253**
power supply, **144**, **254**
pressure roller, 278, **278**, **314**
preventive maintenance (*see* maintenance)
printed circuit board (PCB), **142**, 145, 241
    cleaning, 150, 151-152, **151**

**377**

printed circuit board (PCB), *continued*
  connector on, **258**
  soldering, **343**
programmable VCRs, 13
programming
  bar-code, 20
  on-screen, 20
  recording, 13
pulse code modulation (PCM), 69
pulse width modulation (PWM), 38

## R

random access memory (RAM), 63
real-time counter, 67
RECORD, 14, 15, 71
  not working, 312-313, **313**
recording
  audio signal, 46
  chrominance conversion for, 44
  circuit diagram, **43**
  control-track, 47
  depth multiplex, 58
  luminance conversion for, 43-44
  monophonic/stereophonic linear sound
    tracks, **46**
  programmable VCRs and, 13
  signal conversion for, 43
recording speeds, 4
reel spindles
  adjusting, 242-243, **242**
  cleaning, 163-165, **270**
  height adjustment, 246
remote control, 14, **78**
  angling, 304, **304**
  broken, 194-195
  cleaning, 167-171, **191**
    battery department, 168, **169**
    exterior, 168
    interior, 169, **170**
  interference, 213-214
  location for, 77
  not working, 302-304
  replacement, 98-99
  testing, 170-171
  third-party, 99
  UHF, 99
  universal, 98
repair technicians, 233-234
repairing (*see* cleaning; maintenance; servicing; troubleshooting)
resistance, 102, 105
review, 16, 71

REWIND, 13, 14, 15, 70, 216
  won't operate, 279-281
RF modulator, 95, **144**
RF signal, 29-30
RF switchers, 93-95
routing switchers, 93-95

## S

safety
  during cleaning, 135-136
  volt-ohmmeter, 113-114
  warning signs on back of VCRs, **136**
servicing (*see also* cleaning; maintenance,
  troubleshooting)
  attaching F connector, 345-347, **346**
  audio head adjustment, 244, **244**, **245**
  camcorders, 321-328
  costs, xv
  damaged tapes, 195-196
  dropped VCR, 183-187
  erase-head tape post adjustment, 245
  fire-damaged VCR, 187-189
  foreign objects in VCRs, 192-193
  function switch and control motor/sole-
    noid block diagram, **281**
  fuses, 250-251, **251**
  leaked batteries, 193-194
  manuals and, 240-241
  miscellaneous adjustment procedures, 246
  non-VCR problems, 197-225
  reel spindle adjustment, 242-243, **242**
  remote control units, 194-195
  removing top cover, **141**
  repairing it yourself, 239-240
  sand/dirt/dust in VCRs, 191-192
  screw locations on back of VCRs, **139**
  smoke-damaged VCR, 188-189
  soldering tips/techniques, 341-344
  supplies, 125-129
  test equipment for, 110-123, 235
  tools for, 110, 183, 235, 341-342
  top view of VCR without cover, **142**
  water-damaged VCR, 189-191
  workspace area, 109-110
shaft encoder, **316**
signal amplifiers, 84, **84**
signal splitter, 81-82
signals
  audio, 46
  baseband, 29-30
  color burst, 66
  color subcarrier, 27

luminance, 27
RF, 29-30
television, 26-27
skin effect, 103
slow motion, 15, 63, 64
solder, 342
solder vacuum, 342
soldering, 341-344
  basic, 342-343
  electrostatic discharge and, 344
  good solder joints, 344
  replacing components, 343-344
  technique, **343**
soldering iron, 341
  cleaning and maintenance, 344
soldering tip, 342
sound (*see also* audio)
  obtaining better, 89-90
  poor, 299-302
  speakers "sweet spot," **90**
  troubleshooting problems with, 297-299
sources, 329-335
special effects, 15-16
speed
  capstan, 39
  fast scan, 16
  linear tape, 9
  playback, 3-4
  recording, 4
  slow motion, 15, 63, 64
  writing, 9
spray cleaner, 125-126
standard play (SP), 3
static electricity, 140, 344
stereo
  linear, 10
  synthesizers/decoders, 98
stereo adaptable, 16
stereo audio heads, 9-10
stereo ready, 16
stereo receiver, hooking VCR to, 89-90, **89**
Super Beta, 18-19, 59
super long play (SLP), 3
Super VHS, 18, 59-62
  signal layout, **60**
supply reel spindle, 41, **40**
sync stabilizers, 97
synchronization, horizontal/vertical, 27

**T**

take-up reel, **177**
take-up spindle, 37, 40-41, **40**

tape counter, doesn't operate properly, 315-317
tape loading mechanism, 266, **266**
tape threading, 32-37
  Beta, **33**, **34**, 35-36
  unthreading, 37
  VHS, 33-35, **35**, **36**
tape transport, 37-41
tape-guide spindle (*see* reel spindles)
tapes, 90-92, 214-215
  anticopying process, 97, 217-225
  audio, **4**
  back-tension tape adjustment, 246
  bad rewinder, 216-217
  Beta, **3**
  Beta cover release pin, **172**
  cassette-in leaf switch adjustment, 246, **246**
  cleaning end-of-tape sensor, 162-163, **163**
  damaged, 92-93, 195-196, 214-215
  damaged shells, 214
  handling, 90-92
  head cleaning, **153**
  jammed, 172-173
  life span, 91-92
  loading-end leaf switch adjustment, 246, **246**
  old, 171
  ratchet and pawl mechanism, **91**
  scalloped, 174
  shell damage, 175-177, **176**
  splicing, 174-175, **175**, **215**
  storing, 90-91, 177
  stretched, 171-172, **215**
  using test, 130-131
  VCR eats, 273-278
  VHS, **3**
    cover release pin, **173**, **263**
    reel unlock button, **173**
  video signal frequency recorded on, **45**
  warping, 90
  will not eject, 265-268
  will not load, 260-264
  will not thread, 268-270
  won't play, 270-273
television, 23-29
  adding color, 29
  brightness control, 198
  broadcast spectrum, 26, **26**
  channels, 197
  contrast control, 198
  definition/description, 25-29

**379**

television, *continued*
  degaussing the screen, 101
  electron beam of TV tube, **26**, **29**
  fine-tuning control, 197
  flickering picture, 198-199
  frequency spectrum, 349-353
  history, 23-25
  interference, 202-211
  Macrovision and, 221
  multiple with VCR hookup, 93-95
  picture frames, 28-29
  picture tubes, 27-28
  program reception, 200-201
  reception, 198
  signals, 26-27, **28**
  switch selector, 197
  used for servicing VCRs, 125
  vertical hold control, 198
terminals, input/output, 15, **192**
test equipment, 110-123
  frequency meter, 119-120, **120**
  infrared light detector, 120-123, **122**, **124**
  logic probe, 114-117, **115**, **116**, **316**, **317**
  logic pulser, 117, **117**
  oscilloscope, 118-119
  volt-ohmmeter (*see* volt-ohmmeter)
testing
  ac leakage, **309**
  cables, **138**
  coil and terminals of relay, **252**
  integrity of power cord, **253**
  leakage current, 186-187, **186**, **187**
  motor, 264, **264**
  operation of voltage regulator, **255**
  VCR hookup, 86-87
  voltage levels, **317**
threading, tape (*see* tape threading)
TIME, 71
time shifting, 13
time-base corrector (TBC), 64-67
time-base error, 65
TIMER, not operating properly, 290-292
tires, maintenance, 146-149
tools, 110, 183, 235, 341-342
  screwdriver blade tips, **243**
TRACKING, 71
  no effect on tape, 283-285
tracking control, 211-212
transformers, 81
transmission clutch, 164-165
transmission system, 164
transport mechanism, **142**, **145**

traps, 83, 207
troubleshooting, 227-319 (*see also* cleaning; maintenance; servicing)
  battery problems, 318-319
  bent/broken coil brackets, 315
  Beta misthreading, 314-315
  cassette will not eject, 265-268
  cassette will not load, 260-264
  color sometimes flashes on/off, 313
  dropped VCRs, 183-187
  essence of, 227-228
  excessive dropouts, 318
  fast forward won't operate, 279-281
  flowchart guidelines, 229-231, 247-248
  front-panel indicators not functioning, 288-290
  installation guide, 88
  interference problem/solution chart, 208-211
  picture flags, 317-318
  remote control does not operate properly, 302-304
  rewind won't operate, 279-281
  search does not work correctly, 281-283
  snowy video; poor audio, 299-302
  sound okay; video not okay, 294-297
  sparkles in playback, 314
  tape counter doesn't work, 315-317
  techniques, 234-237
  timer does not operate properly, 290-292
  tracking control has no effect, 283-285
  tuner channels don't change on TV, 292-294
  VCR does not respond to front-panel controls, 285-288
  VCR does not turn on, 249-255
  VCR eats tape, 273-278
  VCR makes unusual mechanical noises, 305-307
  VCR overheats, 310-312
  VCR turns on but nothing else, 255-260
  VCR will not play tape, 270-273
  VCR will not thread tape, 268-270
  VCR won't record, 312-313
  video okay; sound not okay, 297-299
  weak audio playback, 314
  weak video playback, 314
  what you can repair, 231-232
  what you can't repair, 232-233
  you receive electrical shock when touching VCR, 307-310

tubes
  electron beam from, **26**, **29**
  picture, 27-28
tuner
  cleaning, 161-162
  electronic, **12**
  troubleshooting, 292-294
tuning, 11-13
TV/VIDEO switch, 212-213
twin lead cable, 79

## U

UHF IN, 71
UHF OUT, 72
UL listing, 359-360
ultra high frequency (UHF), 12
UV filters, 326

## V

variable crystal oscillator (VXO), 44
VCPs, 6
VCR Plus+, 20-21
VCRs
  back panel diagram, **73**
  block diagram, 30-31, **31**
  brands/manufacturers of, 355-360
  cable-ready, 12-13
  disassembling, 139-140
  enhanced, 15-21
  formats, 2-5
  history, xv
  installing (*see* installation)
  location/placement, 100-101
  operating controls, 13-14, 70-71, 165-166,
    **165**
  operation, 14-15
  overview, 1-21
  problems with (*see* cleaning; maintenance;
   servicing; troubleshooting)
  programmable, 13
  tuning, 11-13
  types, 5-6
  unpacking, 75-76
  VHS system, **2**
ventilation, 76-77
vertical interval reference (VIR), 220
vertical synchronization, 27
very high frequency (VHF), 12
VHF IN, 71
VHF OUT, 72
video
  accessories, 93-99

baseband, 30, 93-95
  composite, 29
  RF, 93-95
  routing switchers, 93-95
  snowy, 299-302
  stabilizers, 97
  troubleshooting problems with, 294-297
  weak, 314
VIDEO DUB, 71
video dubbing, 16-17
video head drum, 8, **8**, **9**, **142**, **153**
  electrostatic discharge unit on top of, **315**
video heads, 8, **7**, **11**
  8mm system, **70**
  automatic cleaner, 158
  azimuth, 51
  cleaning, 152-159
    cassettes for, 152-154, **153**
    costs, 158
    hard-to-clean, 157
    manual, 154-157, **155**, **156**
  demagnetizing, 162
  dirty, 50-51
  drum, 37-38
  five-head systems, 52, **53**
  gap size, 48-49
  noise bars and, 51-52
  path during fast-scan playback, **52**
  self-cleaning, 21
  six-head systems, 52-53
  three-head systems, 52, **53**
video home system (VHS), 2
  hi-fi audio, 57-58
  hi-fi audio signal layout, **58**
  signal layout, **60**
  Super, 18, 59-62
  tape cover release pin, **173**, **263**
  tape reel unlock button, **173**
  tape threading, 33-35, **35**, **36**
  videocassette, **3**
VIDEO IN, 72
VIDEO OUT, 72
video tracks
  alternating, **51**
  recorded on 8mm tape, **69**
  spacing, 49, **49**
videocassette players (*see* VCPs)
videocassette recorders (*see* VCRs)
videocassette tapes (*see* tapes)
volt-ohmmeter, 110-114, **112**
  accuracy, 112
  analog, 111

volt-ohmmeter, *continued*
   automatic ranging, 111
   checking wire continuity, **236**
   connecting to switch terminals, **235**
   cost, 111
   digital, 111
   functions, 112-113
   safety, 113-114
   supplies, 113
   test points for checking double-pole/dou-
    ble-throw switch, **236**
   testing cables, **138**
   testing coil and terminals of relay, **252**
   testing for ac leakage, **309**
   testing for leakage current, 186-187, **186**,
    **187**
   testing for voltage levels, **317**
   testing integrity of power cord, **253**
   testing operation of voltage regulator, **255**
   testing motor, 264, **264**
   using, 114, 235-237

## W

white clip level enhancer, 18
wires, broken, 185, **185**
writing speeds, 9

## Y

Y-C connectors, 61
Y-C INPUT, 72
Y-C OUT, 72

## Z

Zworykin, Vladimir, 24